The World Trade Organization and the Environment

Also by P. K. Rao

INTERNATIONAL ENVIRONMENTAL LAW AND ECONOMICS

SUSTAINABLE DEVELOPMENT: Economics and Policy

THE ECONOMICS OF GLOBAL CLIMATIC CHANGE

The World Trade Organization and the Environment

P. K. Rao
Director
Center for Development Research
Princeton, New Jersey
USA

First published in Great Britain 2000 by
MACMILLAN PRESS LTD
Houndmills, Basingstoke, Hampshire RG21 6XS and London
Companies and representatives throughout the world

A catalogue record for this book is available from the British Library.

ISBN 0–333–77720–4

 First published in the United States of America 2000 by
ST. MARTIN'S PRESS, LLC,
Scholarly and Reference Division,
175 Fifth Avenue, New York, N.Y. 10010

ISBN 0–333–77720–4

Library of Congress Cataloging-in-Publication Data
Rao, P. K.
 The World Trade Organization and the environment / P.K. Rao.
  p. cm.
 Includes bibliographical references and index.
 ISBN 0–333–77720–4
 1. World Trade Organization. 2. Free trade. 3. Sustainable development.
 I. Title.

HF1385 .R36 2000
333.7—dc21
                                                          00–033265

This book is printed on paper suitable for recycling and made from fully managed and sustained
forest sources.

10   9   8   7   6   5   4   3   2   1
09 08 07 06 05 04 03 02 01 00

Printed and bound in Great Britain by
Antony Rowe Ltd, Chippenham, Wiltshire

For my parents

Contents

List of Figures

List of Boxes

Preface and Acknowledgements

The current economic paradigm supporting freer trade and multilateralism needs to be viewed in a broader context of trade-economic growth-environment and sustainable development. In the absence of such an integration the free trade phenomena are unlikely to be sustainable. The externalities of global trade expansion include the impact of trade on the environment. This impact has temporal and spatial variations, with the implications for a delicate balancing of some of the conflicting objectives. The greater the awareness of the implicit and explicit tradeoffs in this regard, the greater will be the potential for sustainable free trade phenomena. Subject to certain conditions, trade expansion tends to enhance the technical and financial resource potential to address environmental problems attributable to both trade and non-trade effects. The key issue here is whether such potential will, in fact, be properly tapped to mitigate the emergence of adverse environmental features and phenomena (resulting from anthropogenic or other influences).

The World Trade Organization (WTO) is an eminently qualified institution to achieve both the objectives of promoting multilateral trade as well as enhancing the quality of the environment. The complementarity of trade and the environment was recognized (albeit partially) even at the early stage of launching this new institution. The preamble to the charter of the WTO clarifies this position in least ambiguous terms. However, it is by no means an easy task to evolve a consensus among all the member countries regarding some of the specific operational norms for conducting multilateral trade activities in order to achieve the agreed upon objectives.

This book augments some of the knowledge regarding the policy and practice aspects of world trade so as to achieve the complementarity of trade and the environment. After a critical review of some of the conventional trade theories, later chapters address the issues of integration of trade and environmental policies, and provide a pragmatic perspective. This perspective is expected to be useful for the WTO and its member (and potential member) countries and various international/national institutions. Much of the discussion and analysis is presented in a largely nontechnical fashion to enable wider readership. Policy-makers, professionals in economics, environment and public policy, in addition to researchers and graduate students in related disciplines are expected to find this book useful in the advancement of understanding of the critical issues of trade and environment.

I am grateful to the World Trade Organization for its permission to quote several of its documents. I am also pleased to acknowledge the support of Ms Alison Howson, commissioning editor, in bringing this book to its fruition. Editorial consultant Mr Keith Povey provided an excellent review. I wish to thank him and the production/copy-editors as well as the marketing staff for their helpful role in this book's publication. My daughters Uma and Usha, and my wife Prema have been constant supporters of my deep involvement with book writing activities.

Princeton, New Jersey   P. K. RAO

List of Abbreviations

AFTA	ASEAN Free Trade Association
ANCOM	Andean Common Market
ANZCERTA	Australian-New Zealand Closer Economic Relations Trade Agreement
APEC	Asia-Pacific Economic Cooperation
ASEAN	Association of Southeast Asian Nations
BECA	Border Environment Cooperation Agreement
BECC	Border Environmental Cooperation Committee
CAFE	Corporate Average Fuel Economy Regulation
CARICOM	Caribbean Community
CBD	Convention on Biological Diversity
CE	Cost Externalization
CFCs	Chlorofluorocarbons
CITES	Convention on International Trade in Endangered Species
CM	Common Market
COMESA	Common Market for Eastern and Southern Africa
CTE	Committee on Trade and Environment
CU	Customs Union
CUSFTA	Canada–US Free Trade Agreement
CVD	Countervailing Duty
DPG	Domestically Prohibited Good
DSB	Dispute Settlement Body
DSU	Dispute Settlement Understanding
EC	European Commission
EEA	European Economic Area
EFTA	European Free Trade Association
EMIT	Environmental Measures and International Trade
EPA	Environmental Protection Agency
ESA	Endangered Species Act
ETM	Environmental Trade Measure
ETP	Eastern Tropical Pacific
EU	European Union
FAO	Food and Agriculture Organization

FDI	Foreign Direct Investment
FFB	Fresh Fruit Bunches
FPET	Factor Price Equalization Theorem
FTA	Free Trade Area
FTAA	Free Trade Area of the Americas
GATS	General Agreement on Trade in Services
GATT	General Agreement on Tariffs and Trade
GCC	Gulf Cooperation Council
GDP	Gross Domestic Product
GPC	GATT Participating Country
GSP	Generalized System of Preferences
HO	Heckscher-Ohlin
IBRD	International Bank for Reconstruction and Development
ICREA	International Commodity-related Environmental Agreement
IDA	International Development Association
IMF	International Monetary Fund
IPPC	Intergovernmental Panel on Climate Change
IPR	Intellectual Property Rights
ITO	International Trade Organization
LDC	Less developed country
LRTAP	Long-range Transboundary Air Pollution
MEA	Multilateral Environmental Agreement
MERCOSUR	Mercado Común del sur (Mercos South Cone Market)
MFN	Most Favoured Nation
MMPA	Marine Mammal Protection Act
MTN	Multilateral Trade Negotiations
NAAEC	North American Agreement on Environmental Cooperation
NAFTA	North American Free Trade Agreement
NGO	Non-Governmental Organization
NNP	Net National Product
NTB	Non-Tariff Trade Barrier
ODS	Ozone-depleting Substances
OECD	Organization for Economic Cooperation and Development
PFTA	Preferential Free Trade Arrangement
PIC	Prior Informed Consent

POP	Persistent Organic Pollutants
PP	Precautionary Principle
PPM	Process and Production Method
PPP	Polluter Pays Principle
PTA	Preferential Trading Arrangement
R&D	Research and Development
ROO	Rules of Origin
RTA	Regional Trading Arrangement
SADC	Southern African Development Community
SCM	Subsidies and Countervailing Measures
SD	Sustainable Development
SPPP	Strong Polluter Pays Principle
SPS	Sanitary and Phytosanitary
SSD	Strong Sustainable Development
STC	Scale, Technique and Composition
TBT	Technical Barriers to Trade
TED	Turtle Excluder Device
TPRB	Trade Policy Review Board of the WTO
TPRM	Trade Policy Review Mechanism
TREM	Trade-related Environmental Measure
TRIM	Trade-related Investment Measure
TRIPS	Trade-related Aspects of Intellectual Property Rights
UDEAC	Central African Customs and Economic Union
UNCED	United Nations Conference on Environment and Development
UNCTAD	United Nations Conference on Trade and Development
UNECE	United Nations Economic Commission for Europe
UNEP	United Nations Environment Programme
VER	Voluntary Export Restraint
WCED	World Conference on Environment and Development
WTO	World Trade Organization
WPPP	Weak Polluter Pays Principle

Part I
Background

1
Global Trade Regimes

1.1 Introduction

A functioning international economy is a prerequisite for continued economic and social advancement. Still, such an economy may or may not be at its efficient or 'optimal' levels of operation. A mere functioning economy is one of several conditions necessary for a sustainable economy and sustainable development (SD); such a functioning economy cannot lead to a sufficiency requirement for SD. A constellation of economic institutions and operations which are oriented toward these goals at the global, regional, national and local levels are necessary in order to put SD in place. International trade and finance are the two foremost significant aspects of the international economic linkages. Whether or not deeper economic integration is important for the international economy is not the main issue. Rather, it should be seen as a derivative of the objective of global welfare maximization: enhancement of the wealth and welfare of the current and future generations. The role of trade in this context remains important. In order that trade remains sustainable, the imperatives of sustainable development must be incorporated for sustainable trade regimes and policies. This holds well in the context of the popular attempts toward trade liberalization. The congruity as well as conflict in free trade and environmental protection need to be elucidated. This book focuses on these issues within the context of the role and activities of the new global apex trade body, the World Trade Organization (WTO).

The long history and tradition of inter-regional trade is abound with multiple and varied experiences during the past few thousand years. Some were largely for mutual economic gain, some were aimed at exploitation, and a few others led to colonization and alien rule of many regions of the world. The costs of enforcement of 'free trade' or legitimate trade as well as containment of illegal trade has been continually on the increase. This is partly due to vastly enhanced volume of global trade, and also due to increased role of individual preferences and consumption patterns that may or may not remain consistent with the domestic and international trade laws. This scenario brings to light some of the limitations of free market approaches to trade and consumption, from an enforcement perspective. The role of market and non-market institutions in governing the global trade remains an issue of perennial importance, and any pragmatic approaches tend to provide changing guidelines which accommodate the dynamics of the economies, institutions and the environmental factors for sustainable development. The volume of trade at the end of the twentieth century relative to that at the middle of the century went up by 1,600 per cent, depicting a linear growth rate exceeding 30 per cent per annum. The role of reduced tariff and non-tariff barriers to trade, and the contribution of the global trading arrangements under the General Agreement on Tariffs and Trade (GATT) have been significant contributing factors in this process.

The Uruguay Round of Trade Negotiations (1986–94) under the GATT framework, the last of the different Rounds of negotiations, alone is estimated to have reduced the tariffs by about $200 billion per annum. Several studies established the positive links between trade liberalization, openness of the national economies, global trade regime effectiveness and economic growth rates of the national economies. However, whether this economic growth is sustainable, and whether it affects the environmental and ecological resource base to such an extent that the trade expansion could be detrimental to long-term growth, remain relevant issues for further investigation.

It is not assumed here that expanded trade is necessarily harmful for the environment. If it were so, we would not even be able to trade goods and services of the 'environmental industry' to protect and upgrade the quality of the environment at local, regional, and global levels. Also, it is not presumed here that trade in all forms under all trading regimes is desirable, nor that trade does not hurt the environmental quality. The fundamental premise here is that trade does affect the environment, sometimes beneficially and on other occasions harmfully. The main focus is to decipher the relevant factors that affect these outcomes, the roles of domestic and international institutions in influencing or steering the outcomes in a desired direction so as to ensure sustainable development. Since each measure requires a series of institutional arrangements, these are evaluated within the framework of the global trading regime with the nodal oversight of the WTO. The concern of this book is centred around this issue. Now that the member countries in the 135-strong WTO agreed to abide by its charter and governance of international trading activities, we will examine the issues within that framework with an analysis toward suggesting pragmatic policies and reforms which augment environmental quality while encouraging trade liberalization. Thus, the desirable objective is to explore 'win–win' policy regimes and evolve relevant guidelines which may be used whenever deviations in favour of the 'second-best' or other relatively sub-optimal policies are inescapable for accommodating other real world considerations. In other words, if a compromise is essential, what constitutes an 'efficient compromise' is also an important practical issue. It is proposed to examine these considerations, without necessarily suggesting that the primary objectives and requirements of sustainable development are to be traded away with any and every feeble pretext or excuse.

This chapter provides a brief historical background to the developments in economic thought as well as in the evolution of institutions in the post-World War II era. The arguments which seek to support and oppose 'free trade' are summarized, as are also the roles and limitations of some of the most popular economic theories in this context. The significance of trade liberalization and its wider economic–ecological–environmental implications are partly deliberated, as a prelude to the detailed arguments given in Chapter 2. Some of the important economic foundations of trade theory are critically reviewed to revaluate the same for current real world configurations. Later, the history and contribution of the GATT, and the background to the creation of the relatively new WTO are proposed for two of the sections in this chapter. Since the emerging global trading patterns are not necessarily under the strict category of 'free trade' regimes, the diagnosis of the underlying factors, which might make it imperative to seek rather 'second-best' trading arrange-ments, remains important for any pragmatic policy interpretation. These aspects are briefly summarized later. The emergence of several regional trading arrangements (RTAs) and trading blocs in recent years poses major new challenges to the global trading arrange-ments; some of these issues are highlighted later in this chapter.

1.2 Free trade, protectionism and managed trade

We need to examine some of the historical as well as fundamental concepts that laid foundations to current perspectives on the eco-nomic and management principles of international trade. This intro-spection, reflecting on the explicit and implicit assumptions of idealized trade models enables one to recognize a few realistic fea-tures: excessive obsession with free trade or some of the assumed applicability of the traditionally known 'theory of comparative advantage' may not always be appropriate.

A parable

If we start with some of the assertions of Adam Smith in 1776, we might still be able to make a beginning and that only. Let us recall the parable of the beaver and the deer in the hunters' world: if among a nation of hunters, for example, it usually costs twice the labour to kill a beaver that it does to kill a deer, one beaver should naturally be exchanged for or be worth two deer (Smith, 1776: p. 49). Smith took his measure of real price, as distinct from nomi-nal price, in labour terms. David Ricardo (1821) sought to assess the real price by the quantity of labour expended. Ricardo's theory of comparative advantage has been an appealing argument favouring expanded trade activities during several years that followed his enunciation of the principle. This theory presumes the existence of situations in which capital and labour do not flow freely between nations, and in which there exists a difference in the wage rates and net private return on capital. It was also suggested, considering the primarily agrarian nature of most economies at that time, that the critical determinant of rise in prosperity for the land owners and the labour is critically dependent on the soil fertility factor. Ricardo recognized that free trade among countries will maximize global welfare, taking all 'climatic' differences as given. The reference to

climate here is not necessarily the meteorological or atmospheric aspects, but these are not excluded either. The original reference to various factors was a general one, perhaps to include technology as we understand it now and to other natural differences across various regions. The Ricardian model lets comparative advantage be critically influenced by modes of production or differences in technology across different trading entities. This was considered usually as a sufficient, but not a necessary, condition for comparative advantage. Some of the recent empirical studies (especially Trefler, 1995; Davis *et al.*, 1997) have established that technological differences are usually a necessary requirement to explain the role of comparative advantage principles in international trading patterns in the recent years of the twentieth century. Later, we shall be able to explore more on the implications of exploiting environmental differences within the context of the applicability and limitations of the theory of comparative advantage.

Eli Heckscher (1919) and Bertil Ohlin (1924) proposed a general economic reasoning to the phenomena of international trade and patterns of trade specialization. Paul Samuelson (1948) strengthened the approach with an analytical basis and detailed analysis using a two-factor, two-sector version of the Heckscher–Ohlin (HO) theorem. According to this comparative advantage principle, a country tends to export a product that is relatively intensive in the factor of relative abundance in that country compared to the importer country.

The Heckscher–Ohlin (HO) factor price equalization theorem (FPET) states the following (see, for details, a classical text like Kemp, 1964): suppose that each of two countries produces something of the same two products with the same two factors, that returns to scale are constant and returns to proportions diminishing, that trade between two countries is freely allowed with no impediments. Then the same real and relative factor prices prevail in each country.

It is also important to recognize that this theorem was concerned with factor rentals and not prices of durable capital assets, including environmental assets.

This theorem, which is also referred to as the Heckscher–Ohlin–Samuelson theorem, is important not only because it focuses attention on the obstacles to equalization of returns to factors of production. This theorem assumes that identical technology is available even in unequal trading partners. In general, the HO theorem is supply based: it takes production technology and resource endowment as a given and tends to act as the main determinant of international trade flows. An implicit assumption in the HO theorem, brought to light by Leamer (1984), relates to the role of the balance of trade (or foreign exchange needs of modern developing economies or other transitional economies). Leamer restated the HO theorem as follows (p. 8):

> A country with balanced trade will export the commodity that uses intensively its relative abundant factor and will import the commodity that uses intensively its relatively scarce factor.

Some of the specific assumptions of the HO formulations include the following restrictive ones (see also Burtless, 1995; and Wan and Long, 1998):

A. Production technology and nature of production
   Production is separable by products to fair degree of disaggregation
   All outputs are internationally tradable
   Constant returns to scale prevail
   Production lags or other gestation lags do not exist
   The production functions are such that the relative factor intensities remain the same irrespective of the factor prices
   Trade and production adjustments to changing influences do not entail significant costs to the enterprises or to the society
   Transaction costs are zero or negligible
B. Production factors
   Factors of production do not enter the utility or welfare function
   Factors are primary and not produced
   Factors are not marketable across borders
   Factor markets are free from distortions
   Factor supplies are not constrained over time
   Factor mobility across sectors does not entail transaction costs

Also, as explained in Burtless (1995), this presumes the invalidity of the phenomenon of factor intensity reversals over time. The latter occurs between two countries when one produces a good using, say, a relatively labour-intensive technique whereas the other produces using a relatively environmental resource-intensive technique. A detailed critical examination of the analytical and empirical limitations of the HO theorem can be seen in Horvat (1999). For a comprehensive presentation of international trade theory, see for example, Gandolfo (1998).

The relative scarcity of environmental endowments in a country is affected by the following (see also Siebert, 1987): the characteristics of sources and sinks of the environmental goods and 'bads', especially the assimilative or renewal capacity of natural resources and environmental pollutants; the environmental preferences – whether directly in terms of the components of the environmental consumption basket (like clean air and water) or in terms of the risk-taking behaviour toward the adverse impacts of environmental bads (like affliction with ill-health and loss of enjoyment of life). The relative roles of the state and the society in this context are distinguishable and these affect the choice of production, technology, consumption and trade patterns.

Some of these trade theories were advanced by respected economists decades before there was sufficient awareness about the environmental limitations of the economies in general. Thus, the factors of production, although logically admit consideration of the environmental inputs in the cost considerations or as factors of production in general, the insistence on comparison of relative costs makes it very difficult to utilize this approach as such in incorporating the environmental factors properly. An obvious interpretation using this approach is to internalize the environmental costs, i.e. add the costs of production and costs of loss of environmental assets and quality (leaving aside the issues of the global and local dimensions in this regard) and then apply the traditional theories. A variant of the FPET can still be derived where the factors need not be all evaluated at market prices but their valuation is done based on imputed values or contingent valuations. These obvious extensions deviate from the original spirit of the arguments by incorporating adjustments that are not perceived in the market place as some of the transparent costs. We, then, are forced to deviate from the traditional theories and their underlying market assumptions. This is by no means an attempt to ridicule the role of market institutions but an appreciation of applicable limits helps.

All that this amounts to is to assert that the real costs of the goods or products is usually understood by the prices reflected in the trading processes and that major changes in the advocacy of the theory of free trade might be called for. Again, this is not an indictment of the free trade theory in its entirety. Rather, in the absence of proper recognition of the role of inputs that are assumed free but will eventually be either exhausted or very expensive, free trade theory is bound to hit its physical, ecological, environmental and economic barriers. A beginning can be made by incorporating the non-marketized inputs explicitly in the meta-production functions, and recognizing the interrelationships between traditional inputs and these augmented inputs. Such an interdependence evolves itself into a dynamical relation and brings about changes in the changing equilibria of the trading optima and price relations. This is predicated on the assumption that this process is being governed by a multilateral approach to the issues and that there is no free rider environmentalism: some pay for others' free access to sustained or enhanced global environmental output. In essence, the role of missing markets must be recognized if sustainable trading patterns and sustainable development is of concern. Yet, there is a great deal of detailed structuring of the approach involved in obtaining prescriptions for a useful policy governing global trade and environment in an integrated manner.

It is relevant to recall that 'free trade is not optimal in imperfectly competitive industries' (Helpman and Krugman, 1992: p.185). This arises from an earlier foundation in economics: zero tariffs is not optimal in free trade (Meade, 1950; Lipsey, 1957). The main reasons for these positions include the limitations of market institutions to fully address negative economic (and environmental) externalities.

Missing trade mysteries

Traditional trade theory suggests the predominant or exclusive role of the relative factor endowments. However, empirical studies do not seem to lend support to this long-maintained premise (see, for example, a detailed survey in Leamer and Levinsohn, 1995). Countries do not trade as much with each other as their endowment differences suggest about such possible trade under comparative advantage principles or FPET. Davis (1997) and Trefler (1995) found that it is not possible to find any significant empirical support to the FPET.

Trade is missing relative to the FPET. Trefler (1995), using the data for 33 countries which accounted for about three-quarters of the world exports suggested modifications which allow the role of habit formation and local production features which affect the domestic consumption and international technology differences; the latter affect the efficacy of production and hence explain the trading patterns better than the traditional factor endowments or their differences. The role of the country size, more specifically the absolute magnitude of skilled labour is found to support the validity of this assertion, as demonstrated empirically in Tortensson (1998).

The fact remains that it is rather futile to seek a static explanation for a dynamic phenomenon where the structure and composition of international trade is governed by a set of endogenous factors like the dynamics of technological change and institutional change. 'We need a more technologically oriented trade theory and more emphasis on dynamics', argued Helpman (1998) using a rather detailed survey of the analytical and empirical issues.

A number of elements of transaction costs are noteworthy. Some of these include the following: the trading entities' perception of counterparts' trade terms and their legal enforceability, and the national governments' explicit and implicit policies toward trade with specific countries and governing specific products for export or import, the role of effective taxation on each of the trading activities and their aggregative contribution to the enterpreneur's profitability, existence of transparent trade laws and enforcement methods – especially the customs procedures and the role of corruption in enforcement levels. These elements affect the potential trade transactions, and these are not usually independent of the national differentiation or geographic reference. Thus, as Krugman (1991) argued, 'Nations matter … because they have governments whose policies affect the movements of goods and factors. In particular, national boundaries act as boundary to trade and factor mobility'. The national borders tend to affect trade significantly, even under the regimes of 'free trade'. Helliwell (1998) rightly argued that geography and national borders have separate but analogous effects in setting patterns of economic activity, and the determinants of the realized trade include transaction costs and the role of history. Given the 'border effects', it was suggested (Helliwell, 1998), ironically, that international trade theory is more suited for trade explanations within rather than across countries. Historically for the past several centuries, institutional differences (including cultural factors, familiarity of which plays the economic role of information and trust), the role of uncertainty and transaction costs remain some of the most robust determinants of trading patterns; for a detailed description see Greif (1992).

The only valid argument for trade protection as a means of maximizing economic welfare is the optimum tariff argument, in addition to any serious practical considerations (Johnson, 1965). Mobility of factors between occupations and territories remains only moderately flexible, the market prices are influenced by taxes and implicit subsidies as well as institutional influences. Consequently these do not necessarily reflect the true social marginal values of the society's output (see, for details, Macbean and Snowden, 1983). The ideal requirements of social optimum may not be attained under free trade but there is no greater possibility of attaining them under any distorted measures either. The issue is which of these or other alternative forms of governance tend to enhance economic welfare in a practical sense. This implies testing the trade regimes against the metrics: relative gains for exporting and importing countries under each trading arrangement or system (like trade liberalization, tariff reduction, reciprocal and multilateral trading arrangements), the corresponding implications for the enterprises involving in the trading activities (from private sector, sector, or both), short-term and long-term economic growth and income distribution implications for trading countries, environmental sustainability as a consequence of expanded trade, and implications for SD. Deadweight loss under different trading restrictions, static and dynamic efficiency as well as welfare aspects at consumer, producer and general levels are some of the features which merit serious consideration in this context.

Are there limits to free trade, even outside the environmental considerations? Perhaps there are a few worth noting. These can be interpreted more against 'free trade' and in favour of some form of managed or strategic trade. The role of irreversibility contributed by expanded trade is sometimes significant. The lack of flexibility and trade in a less diversified economy could potentially foreclose some of the production and investment options that might be available in a diversified trade situation. This is particularly bothersome when the trade involves rather significantly unequal partners, and more so when high discounting of the future on the part of the smaller participant. In other words, if short-term considerations weigh substantially more than sustainability of the trade regimes and the economic development requirements of the domestic economies, a failure of the trade objectives and the sustainability of trade transactions might well be a predictable outcome of such lopsided policies.

If anticipations or expectations have a role to play in trade dynamics over time between different countries, it is possible that a small or less diversified country could stand to become a net loser in its trade transactions with a large or diversified trading partner country as a result of 'free trade'. This is possible as a result of potential production investment irreversibilities or the evolution of capital asset formation as a result of trade expansion subject to volatile trading regimes between the participating countries (and their trading blocs wherever applicable). The key to such potential losses to a small or less diversified country is founded on the likely consequential loss in the bargaining power of the country in relation to the large or diversified country, after trade dependence has been established both in the stock and flow terms. This is caused by the creation of immovable assets and production infrastructure with irreversible investments, and the imperatives to balance current budgets with (uneconomical) exports. Some of these issues are addressed descriptively by McLaren, 1997 in connection with some of the trade problems involving the US and Canada, who argued that 'the rational investments undertaken by small-country citizens... in anticipation of future... can rob the small country of its flexibility'.

Some of the main potential benefits of open or free trade include: enhanced efficiency in production, consumption and competitiveness leading to technical and managerial innovation, as well enhanced foreign investment. However, none of these accrues automatically whenever free trade patterns are followed. These are more likely to occur between somewhat comparable partners. In general, expanded trade tends to provide additional avenues for resource augmentation with the enlargement of the set of opportunities for economic gains. This is not a universally valid assertion, however. Box 1.1 details of various alternative categories of rather canonical forms of trading methods are detailed for clarity of their usage during this and later chapters. It is not difficult to postulate a set of axioms and describe the alternatives in relation to the same. While such an approach

Box 1.1 Alternative trade policies

Open Trade: The trade policy of a country or a group of countries where all legitimate goods and services can be imported and exported, with or without tariffs.

Free Trade: The trade policy of a country or a group of countries which impose no tariff or non-tariff restrictions on the trade of legitimate goods and services; the trade transactions occur purely as a result of demand and supply forces; the government's role in augmenting transaction costs remains at zero level.

Fair Trade: The trade policy which recognizes the role of unequal market powers of trading partners and seeks to find equitable trading arrangements to ensure that the powerful trading partners do not take undue advantage of the relatively less powerful partners.

Balanced Trade: The trade mechanism of a country which seeks to allow export–import operations with due regard to broad-based economic development strategy combined with socioeconomic considerations. Also, the trading methods are not influenced by the compulsions of foreign debt or mobilization of foreign exchange reserves for the national account.

Optimal Trade: The trading policies of a given country or group of countries which seek to optimize a set of financial and socio-economic objectives, taking into account their current and future resources, as well as economic and social factors; the roles of the market influences are partially augmented to achieve desired objectives; this approach presumes a great deal of information availability and processing on the part of a central agency which steers the systems of trade for the country.

Protected Trade: Trade policy of a country or a group of countries which adopt tariff and/or non-tariff trade barriers affecting the transactions resulting from the market forces in order to protect domestic industry, or foreign exchange reserves, or retain a closed economy to perpetuate undisturbed state control of resources and institutions.

Managed Trade: Trade policy which allows direct or indirect state intervention in order to steer the inflow and outflow of goods and services or maintain a level of foreign exchange reserves

Box 1.1 continued

or achieve a set of desired objectives with trade interventions; these interventions coexist with the role of the market and could possibly play a catalytic role.

Sustainable Trade: Trade policy that takes into account not only short-term benefits and costs but also seeks to ensure that trade operations can be sustained for very long time horizons (or for ever) without interruptions caused as a result of trade activities themselves; any potential adverse contributions of trade activities to environment and negative feedbacks to the economy thereof are taken into account. The existence and the role of a responsible governing entity is presumed here, since markets themselves do not usually contain the information and foresight needed to accomplish the integration of (a) economic and environmental considerations, and, (b) factors and impacts associated with markets and missing markets associated with various resources, products and services (including services of nature). Another alternative brief definition of Sustainable Trade is the derived one: the trade mechanism which ensures sustainable development (the latter defined according to rather standard pragmatic norms).

promises greater rigour and precision, the usability in the context of the main concern of this book, is likely to be limited.

Historically, soon after the end of the World War II, a series of multilateral trade reforms were initiated. These sought to reduce a number of tariff and non-tariff trade barriers and principles of non-discrimination among different participating countries under GATT. These details follow in the next section.

1.3 GATT

Multilateral approaches to trade policy

The nature of the evolution of the global trade regimes and international trade policies has been contributed by an active participation of most countries towards a multilateralism which seeks to accommodate conflicting as well as complementary activities in trade and finance. An institutional framework, however imperfect, was a necessary requirement in this process. This was recognized and attended to as soon as peace returned after the end of World War II.

GATT was negotiated at Geneva in 1947 about the same time when a draft of the charter for the proposed International Trade Organization (ITO) was being prepared by the US and some of its allies. The US Congress failed to approve the 1948 ITO charter and thus the ITO never came into existence. For reasons of the US Constitutional law, the President endorsed the GATT without seeking the approval of the US Congress. The GATT was adopted as a provisional charter, under a Protocol of Provisional Application, expecting the ITO to come into existence. The latter never materialized and the GATT remained a provisional set up right from its formation till its end in 1994. This in itself is creditable that an organization could contribute even as much as it did when it was not a Treaty under the UN Charter, and the parties were simply 'contracting parties' and not members of a charter organization as is now the case with the WTO. Appendix I to this chapter provides major highlights of the history of evolution of multilateral trading regimes.

The first operative objective of the GATT was to facilitate trade expansion via tariff reduction. The significant contribution of the GATT has been via its several multi-year rounds of trade negotiations starting in 1947 and ending in 1994, when the final round (the Uruguay Round) led to abolition of the GATT, replacing it with a more powerful and purposeful world body to govern trade and its liberalization. The initial rounds of GATT deliberations started with the participation of only about 23 countries, and finally led to the participation of 123 countries by the end of the Uruguay Round of Trade Negotiations which ended in 1994 after the Marrakesh Declaration.

It was rather strange that the GATT remained a provisional agreement between contracting parties, rather than a Treaty between member countries. Similarly, the mechanism for enforceability of various Articles and dispute resolution was very feeble. Some sought to describe GATT as the General Agreement to Talk and Talk, and some sought to gut the GATT. The regional level policy contribution of the GATT to enhance the welfare of some of the poorer countries remained highly questionable. In its last round of trade negotiations, the Uruguay Round is seen as having worsened the trade terms as far as some of the least developed countries are concerned. Whereas industrialized countries were allowed to maintain quantitative restrictions in areas like agricultural products and textiles, developing countries were asked to improve their market access for the industrial countries. While tariff reductions for industrial goods averaged 38 per cent for imports from all countries of origin, they were less by 4 per cent for imports from developing countries (Demske, 1997). Most of the gains accrue to developed, rather than developing countries (see also Hamilton and Whalley (1995), including references to some of the studies which estimated that the latter may gain only about one-tenth of the total gains). The UN World Economic and Social Survey 1995 (UN, 1995) described the provision of agricultural subsidies as 'inherently favor developed countries over developing ones'. So much for the reduction of global inequalities and the globalization via trade liberalization under the Uruguay Round of the GATT. Considering the limited nature of functions and effectiveness of the GATT, it may be accurate to state that most significant development in the evolution of institutions and trading regimes since the end of World War II in the field of international trade is the creation of the WTO rather than the GATT and its half a century of provisional existence.

The most important provision of the GATT 1947 was the most-favoured-nation (MFN) clause, stated below:

Article I: General Most-Favored-Nation Treatment
With respect to customs duties and charges of any kind imposed on or in connection with importation or exportation or imposed on the international transfer of payments for imports or exports, and with respect to the method of levying such duties and charges, and with respect to all rules and formalities in connection with importation and exportation, and with respect to the application of internal taxes to exported goods, and with respect to all matters referred to in paragraphs 2 and 4 of Article III, any advantage, favor, privilege or immunity granted by any contracting party to any product originating in or destined for any other country shall be accorded immediately and unconditionally to the like product originating in or destined for the territories of all other contracting parties.

This feature was sought to be maintained by all the contracting parties of the GATT, with a few exceptions to incorporate special needs for budget balancing in some of the developing countries, protection of infant industries for a few years, and protection of health of humans, plants and animals. Despite a few cases of use of these escape clauses, the MFN constituted the backbone of the entire multilateral trade negotiation processes for all these years. In a recent contribution, Bagwell and Staiger (1999) offered essentially a static analytical framework to interpret the fundamental principles of GATT: reciprocity and non-discrimination. These principles were seen as complementary in generating efficient multilateral trade operations. The reasoning does not extend to include any consideration of environmental externalities, however.

Some of the lacunae and 'birth defects' of the GATT included (see also Jackson, 1995): ambiguity about the powers of the contracting parties to arrive at some specific decisions, waiver of authority and enforcement in cases of misuse of various provisions or exceptions; murky approaches to dispute resolution and the role of consensus requirement; vagueness about many institutional arrangements; lack of clarity of procedures and decision-making mechanisms leading to protracted focus-free negotiations.

In terms of analytical economic reasoning, developed as early as the 1950s, the contributions of Meade (1955) and Lipsey (1957) remain significant. Among other issues, the problem of determining optimal trade tariffs in such a manner as to maximize global economic welfare under a set of plausible objectives was addressed by these authors. They argued that such welfare is likely to be enhanced if trade tariffs are reduced, and that it may not always be optimal to eliminate such tariffs completely. The reason for such a suggestion is that there is usually need for 'second-best' policies which are capable of accommodating important social objectives or other economic objectives like economic growth maximization with social justice. It was also derived as corollary to this reasoning that optimal tariff levels could be equal to, below or above the prevailing levels in a given sector or country. This reasoning lends useful support to the issue of optimal trade liberalization or 'optimal free trade' which is built on sustainable trade and sustainable development principles. Thus, trade liberalization for its own sake or to serve the lopsided interests of a few is not a desirable end by itself. For similar reasons, and as corollary to such alternative premises, there exists a trade regime which is neither protectionist of the trade nor unprotectionist of the environment. The real issue is to find or devise such a regime which can lead to sustainable trade, the main concern of this book within the context of the WTO charter. There are a number of RTAs which also have a major role in trade activities and consequential environmental implications. Technically, the regional regimes are required to be functioning with due notification to the WTO, but a direct intervention from the latter has limited role. About half the global trade is channelled through these regional arrangements, explaining possibly the major role of 'transaction costs'. These costs include the combination of perceived costs of doing business, familiarity, cultural and political integration, geographic proximity, and other factors. Some of these regional regimes are summarized in the next section.

1.4   New trade regimes: regional and global

There have been significant and continuing changes in the international trade arena during the past few decades. These changes did not necessarily reflect the effects of the GATT or other global institutions. Rather, there have been increasing activities at regional levels seeking deeper economic integration. Some argued that the proliferation of the RTAs was a response or an endogenous institutional evolution process to the limited role and effectiveness of the GATT in the governance of global trading arrangements or a rapid expansion of the same. Various types of regional trading arrangements emerged especially since the mid-1960s. The multitude of these is so extensive that by the time the new world body, the WTO was created on the first day of 1995, almost all the member countries were included in one or more of the regional arrangements. About 45 per cent of the global trade is governed by RTAs. By the end of the twentieth century, a record 179 RTAs have been notified to the WTO about their existence and being still in force. The WTO launched a series of investigations to ensure the compatibility of the RTAs with the spirit of trade liberalization and the WTO charter obligations of its member countries. A WTO Committee on Regionalism has been working on a number of relevant issues to ensure the congruence of the regional and global trading regimes. Appendix II to this chapter illustrates a list of regional trade arrangements, categorized into four major groups: customs unions, common markets, free trade areas and economic cooperation agreements. These are defined below:

- Free Trade Areas (FTAs): trade within the agreed area is duty free, these areas eliminate internal tariffs and NTBs but do not harmonize external barriers; each member can set its own duty rates on imports from non-members.
- Customs Unions (CUs): these remove internal barriers and establish a common external tariff; all members levy the same set of customs duties on imports from non-members.
- Common Markets (CMs): these are CUs in which barriers to the mobility of labour and capital are eliminated.

About half a century ago, Jacob Viner (1950) suggested that preferential free trade arrangements (PFTAs) can have beneficial as well as detrimental effects on economic and trade efficiency features at the global level. By eliminating trade barriers among the members of a trading bloc, PFTAs can create trade and efficiency via production efficiency enhancement, and could also divert trade by expanding the production of less efficient members at the expense of others outside the block who could achieve the same, relatively more efficiently (see also Frankel, 1996 and Lawrence, 1996). Viner's original presentation was centered around evaluation of CUs in terms of relative welfare effects arising from costs of production. Later, Meade (1955) and Lipsey (1957) clarified that the removal of subsidies and tariffs could affect welfare via changes in consumption. Thus, welfare in a free trade area (FTA) member could rise even when it is buying from a higher cost supplier, provided the benefits from more efficient consumption exceed the loss of tariff revenues (Lawrence, 1996).

The critical issue in the RTAs and their potential positive or negative roles is that they need to be assessed largely in terms of their 'trade creating' or 'trade diverting' contribution. In general, it is feasible that the RTAs tend to take advantage of lower transaction costs in conducting trade activities, relative to the same operations at the global level. This is because of the spatial, informational, social, and other transaction cost-minimizing elements of total costs of doing business within specified relatively homogeneous regions. Under the GATT regimes, Article XXIV provides for the formation of CUs and their variants with a few requirements to be fulfilled. Detailed statement of this Article is given in Box 1.2. The CUs formed under

Box 1.2 GATT Article XXIV

Territorial Application-Frontier Traffic-Customs Unions and FTAs

24.3 The provisions of this Agreement shall not be construed to prevent:

(a) Advantages accorded by any contracting party to adjacent countries in order to facilitate frontier traffic;

(b) Advantages accorded to the trade with the Free Territory of Trieste by countries contiguous to that territory, provided that such advantages are not in conflict with the Treaties of Peace arising out of the World War II.

24.4 The contracting parties recognize the desirability of increasing freedom of trade by the development, through voluntary agreements, of closer integration between the economies of the countries parties to such agreements. They also recognize that the purpose of a customs union or of free-trade area should be to facilitate trade between the constituent territories and not to raise barriers to the trade of other contracting parties with such territories.

24.5 Accordingly, the provisions of this Agreement shall not prevent, as between the territories of contracting parties, the formation of a customs union or of a free-trade area or the adoption of an interim agreement necessary for the formation of a customs union or of a free-trade area; Provided that:

(a) With respect to a customs union, or an interim agreement leading to a formation of a customs union, the duties and other regulations of commerce imposed at the institution of any such union or interim agreement in respect of trade with contracting parties not parties to such union or agreement shall not on the whole be higher or more restrictive than the general incidence of the duties and regulations of commerce applicable in the constituent territories prior to the formation of such union or the adoption of such interim agreement, as the case may be;

(b) with respect to free-trade area, or an interim agreement leading to the formation of a free-trade area, the duties and other regulations of commerce maintained in each of the constituent territories and applicable at the formation of such free-trade area or adoption of such interim agreement to the

Box 1.2 continued

trade of the contracting parties shall not be higher or more restrictive than the corresponding duties and other regulations of commerce existing in the same constituent territories prior to the formation of the free-trade area, or interim agreement as the case may be;

(c) any interim agreement referred to in subparagraphs (a) and

(b) shall include a plan and schedule for the formation of such a customs union or of such a free-trade area within a reasonable length of time.

Source: GATT (1994)

this Article must 'not on the whole higher or more restrictive' than before their formation. The FTAs are required not to allow any member to enhance barriers to trade to non-members after joining the FTA. As rightly argued by Jackson (1996), the specifications of the Article XXIV are inadequate for current problems of international trade like those influenced by the Rules of Origin (ROO). The possibility that the emerging RTAs could devise instruments which in effect act like protectionist measures without resorting to changes in tariff or non-tariff trade barriers (NTBs) is a realistic one and this feature must be taken seriously in all global trade regime policy formulations and their governance (see also Lawrence, 1996 for more details). One of the studies of the WTO Secretariat concluded (WTO, 1998a) rather prematurely that regional and multilateral integration arrangements are complements rather than alternatives in the pursuit of more open trade. Bagwell and Staiger (1999) argued, using an analytical framework, that FTAs and CUs pose a 'threat to the efficiency properties of the existing multilateral system'.

According to some, the Article XXIV has been one of the most abused and the ever increasing role of the CUs stands a testimony to this interpretation. Hardly any RTA was ever disapproved by the GATT which was notified of the same from time to time. According to Gibb (1994), 'the weakness and lack of competence of the GATT to influence the behavior of a truly globalized world economy have in part, been responsible for the resurgence of regionalism in the 1990s'.

WTO (1998b) states that the RTAs 'have become the norm in international trading relations, and political impetus to increase their number and scope has anything but diminished'. Some of the recent developments in this direction include the following. An illustrative list of previous RTA formations is given in Appendix III to this chapter.

In Africa, the new Economic and Monetary Union gathering Francophone countries of West Africa plans, among objectives, to bring into effect a common external tariff; in southern Africa, negotiations to create a FTA among the members of the Southern African Development Community (SADC) are ongoing; within this group, the Southern Africa Customs Union (including Botswana, Lesotho, Namibia, South Africa and Swaziland) are renegotiating their customs union's relations. The East African Cooperation member States are continuing their progress towards an eventual economic and monetary union. The African members of the Lome Convention are currently engaged in negotiations on a new Convention.

In the Americas, an agreement has been signed between 34 countries to create a Free Trade Area of the Americas (FTAA) by 2005. In addition, progress towards the completion of a common external tariff among Mercos South Cone Market (MERCOSUR) members has continued. Other agreements involve individual countries in the regions or pre-existing sub-regional groupings (for example, the recent agreements between Mercosur and the Andean Community, and between Chile and Canada).

RTAs are seen (Sager, 1997) as (a) a response to the unwieldy framework for GATT negotiations; (b) geographic proximity and economic gravity considerations; (c) constitute a series of new attempts to trade expansion and economic integration; (d) political considerations in the trade policies and their governance; (e) a supplemental effort to the cumbersome tariff or other trade barrier reducing attempts under the GATT where the decision-making usually is about half a decade more, and the political time horizons find that like several political years (not necessarily measures in terms of calender years).

A detailed study on Regionalism organized by the Carnegie Endowment for International Peace (1997) concluded that both on theoretical economic grounds as well as empirical evaluation of the RTAs at work, there are no unambiguous effects of the regional mechanisms on trade creation or trade diversion. Thus, the RTAs remain a double edged weapon for promotion of trade liberalization (within specified areas) and support of protectionism against non-member countries. Thus in the aggregate it may be a net gain or net loss of trade and economic benefit at the global level, depending on the time period and the corresponding economic features obtaining in such time intervals. In other words, these do not necessarily perform according any predictable results when evaluated at the global level. The Carnegie Report recommended that the RTAs could enhance trade and investment creation with the adoption of the following guidelines, in addition to revision of the provisions under Article XXIV of the GATT: (a) precise compliance criteria for RTAs with particular reference to tariffs, rules of origin, and transparency and enforcement mechanisms for the RTAs, (b) member countries in each of the RTAs harmonize their trade rules, (c) the WTO utilize the institutional and procedural details to promote compatibility between RTAs and the WTO itself, (d) WTO devise a code of admission for any member to join a recognized RTA.

If RTAs are a decentralized market and non-market institutional response to a central apex governance by the GATT or the WTO, then we must recognize the role of informational content of trade and other economic activities which are coordinated by different RTAs according their own charters. In other words, these act like mechanisms for manageable decentralization based on functional considerations: history, economic and political geography, information exchange for trade decisions within the framework and institutional arrangements, similarities in cultural and political factors, and various transaction costs in respect of factors not necessarily reflected in the globally standard product price mechanisms. Whereas transaction costs are major influences in the formation of the RTAs, these are unlikely to remain significant elements of trade and competition in the *ex post* for members within the RTA set up.

The role of economies of scale, and increasing returns to scale, reduced uncertainty in effecting profitability are some of the factors influencing the RTA formation. These are seen as purely economic ingredients, although the role of the political and other institutional factors remains significant as well.

1.5 Concluding observations

Much of traditional economic theory presumes the infinite availability of ecological and environmental assets. This implicit assumption led to the focus on principles of comparative advantage as the single most important determinant of international trade. While this premise still has some merit, the empirical evidence shows that the role of a number of other important elements is rather dominant. These elements include national or geographic borders, transaction costs, institutional configurations like the RTAs, political economy of alternate forms of trade patterns, and the role of multilateral institutions (or endogenous institutional dynamics). Freer trade for its own sake is not a desirable end product; it should be ingrated with broader issues of development, and sustainability.

The role of GATT and its impact during its half-a-century of provisional existence indicates its positive impact in reducing trade distortions and tariff barriers. The emergence of the new WTO with a full pledged charter structure and member participation portends much greater influence and impact on the global trade, economy and the environment. These are the issues to be deliberated in the course of the rest of the text. The foundations of an integrated trade–environment–development emerge from those of sustainable development processes and these constitute the focus of Chapter 2 to follow.

Appendix I
Chronology

1944   Bretton Woods Conference on global economic arrangements
1946   The International Bank for Reconstruction and Development (IBRD) and the International Monetary Fund (IMF) start activities
1947   Drawing up of the institution GATT
1947   First Round of multilateral tariff reductions negotiated in Geneva
1948   Adoption of the ITO charter by the International Conference on Trade and Employment in Havana
1949   Second Round of multilateral tariff negotiations in Annecy
1951   Third Round of multilateral tariff negotiations in Torquay
1956   Fourth Round of multilateral trade negotiations in Geneva
1957   Treaty of Rome establishes European Common Market

1960–61	Fifth Round of tariff negotiations under the GATT: Dillon Round
1964–67	Sixth Round of tariff negotiations – GATT: Kennedy Round
1973–79	Seventh Round negotiations – GATT: Tokyo Round
1986–92	Eighth Round negotiations – GATT: Uruguay Round
1989	Trade Policy Review Mechanism (TPRM) introduced in the GATT after mid-term review of the Uruguay Round
Feb. 1992	Signing of the Treaty of Maastricht
Apr. 1994	Marrakesh Declaration – Final Act of the Uruguay Round resolutions and establishing the WTO
Jan. 1995	Birth of the WTO
Nov. 1996	Formal agreements regarding cooperation of the IBRD and the IMF with the WTO
Dec. 1996	WTO First Ministerial Conference in Singapore
May 1998	WTO Second Ministerial Conference in Geneva
Mar. 1999	WTO High Level Conferences on Trade and Environment, and on Trade and Development in Geneva
Dec. 1999	WTO Third Ministerial Conference in Seattle

Appendix II
GATT 1994

1. The General Agreement on Tariffs and Trade 1994 ('GATT 1994') shall consist of:

(a) the provisions in the General Agreement on Tariffs and Trade, dated 30 October 1947, annexed to the Final Act Adopted at the Conclusion of the Second Session of the Preparatory Committee of the United Nations Conference on Trade and Employment (excluding the Protocol of Provisional Application), as rectified, amended or modified by the terms of legal instruments which have entered into force before the date of entry into force of the WTO agreement;

(b) The provisions of the legal instruments set forth below that have entered into force under the GATT 1947 before the date of entry into force of the WTO Agreement:

(i)   protocols and certifications relating to tariff concessions;
(ii) protocols and accession (excluding the provisions (a) concerning provisional application and withdrawal of provisional application and (b) providing the Part II of GATT 1947 shall be applied provisionally to the fullest extent not inconsistent with legislation existing on the date of the Protocol);
(iii) decisions on waivers granted under Article XXV of GATT 1947 and still in force on the date of entry into force of the WTO Agreement;
(iv) other decisions of the CONTRACTING PARTIES to GATT 1947;

(c) the Understandings set forth below:

  (i) Understanding on the Interpretation of Article II: 1(b) of the General Agreement on Tariffs and Trade 1994;
  (ii) Understanding on the Interpretation of Article XVII of the General Agreement on Tariffs and Trade 1994;
  (iii) Understanding on Balance-of-Payments Provisions of the General Agreement on Tariffs and Trade 1994;
  (iv) Understanding on the Interpretation of Article XXIV of the General Agreement on Tariffs and Trade 1994;
  (v) Understanding the Respect of Waivers of Obligations under the General Agreement on Tariffs and Trade 1994;
  (vi) Understanding on the Interpretation of Article XXVII of the General Agreement on Tariffs and Trade 1994; and

(d) the Marrakesh Protocol to GATT 1994.

Appendix III
Regional trade regimes

Regional trading regimes are broadly classified below into four categories, considering their stated objectives in each group.

CUs

1 EU, launched originally as the European Commission (EC) in 1957 under the Treaty of Rome, and achieved economic and monetary integration by 1999.
2 Andean Common Market (ANCOM), formed as a common market among five south American countries and enhanced itself into a CU in 1995 with trade policies which provide preferential trade among member countries and common external tariff.
3 Central African Customs and Economic Union (UDEAC), formed as a common market into 1964 and upgraded into CU in 1994 with the Treaty that established the Economic and Monetary Community of Central Africa; member countries implement a common external tariff, phased out internal customs duties, share a common currency and a central bank.

Free trade areas

1 European Economic Area (EEA), signed in 1992 and became effective in 1994; free movement of goods, services and capital, establishment of common guidelines governing competition and legal enforcement of trade laws.
2 Australia–New Zealand Closer Economic Relations Trade Agreement (ANZCERTA), signed in 1983 and modified in 1988; eliminates antidumping provisions and harmonizes customs procedures and product standards.

3 Association of Southeast Asian Nations (ASEAN) Free Trade Agreement (AFTA), formed in 1992 with the aim of achieving FTA by 2008, with preferential tariff on manufactured goods and regional cooperation.
4 European Free Trade Association (EFTA), formed in 1960, intended as promotion of free trade among countries that did not join the EC.
5 Canada – U S Free Trade Agreement (CUSFTA), signed in 1988 to liberalize tariffs and trade and to harmonize some of the technical standards.
6 North American Free Trade Agreement (NAFTA), signed in 1992 and become operational in 1994; seeks to liberalize trade and investment flows, and side agreements accompanied covering labour and environmental protection.

Common markets

1 Gulf Cooperation Council (GCC), formed in 1981 to encourage free movement of goods and services among six middle-eastern countries – all oil exporters.
2 Mercos South Cone Market (MERCOSUR), signed as a Treaty in 1991 between some of South American countries, seeks to coordinate free movement of trade in goods and services, elimination of tariff and non tariff barriers to trade, and coordinates macroeconomic policy regimes.
3 Caribbean Community (CARICOM), formed in 1973, reduced external tariffs among member countries.

Economic cooperation

1 Asia–Pacific Economic Cooperation (APEC), this forum was established in 1989 for regional trade and economic cooperation, with greater open trade goals to be attained by 2010.
2 South Asian Association for Regional Cooperation (SAARC), formed in 1985 seeks to extend greater economic and trade cooperation among the member countries Bangladesh, Bhutan, India, Maldives, Nepal, Pakistan and Sri Lanka.

Sources: Derived and revised from information based on Anderson and Blackhurst (ed.) (1993), de Melo and Panagariya (ed.) (1993), and especially Table A.1 of Lawrence (1996).

References

Anderson, K. and R. Blackhurst (ed.) (1993) *Regional Integration and the Global Trading System*, London: Harvester Wheatsheaf for GATT.
Bagwell, K. and R. Staiger (1999) 'An economic theory of GATT', *American Economic Review*, 89.1, 215–48.
Burtless, G. (1995) 'International trade and the rise in earnings inequality', *Journal of Economic Literature*, 33, 800–16.

Carnegie Endowment for International Peace (1997) *Reflections on Regionalism – Report of the Study Group on International Trade*, Washington, DC: Carnegie Endowment.

Davis, D. R., D. E. Weinstein, S. C. Bradford, and K. Shimpo (1997) 'Using international and Japanese regional data to determine when the factor abundance theory of trade works', *American Economic Review*, 87.3, 421–46.

de Melo, J. and A. Panagariya (ed.) (1993) *New Dimensions in Regional Integration*, New York: Cambridge University Press.

Demske, S. (1997) 'Trade liberalization – de facto neocolonialism in West Africa', *The Georgetown Law Journal*, 86.1, 155–80.

Flam, H. and M. J. Flanders (ed.) (1924) *Heckscher-Ohlin Trade Theory*, Cambridge, MA: MIT Press.

Frankel, J. (1996) 'Does regionalism undermine multilateral trade liberalization or support it?', in *Regional Trading Blocs*, ch. 10, Washington, DC: Institute of International Economics.

Gandolfo, G. (1998) *International Trade Theory and Policy*, Heidelberg: Springer Verlag.

GATT (1994) *The Results of Uruguay Round Trade Negotiations*, Geneva: GATT Secretariat.

Gibb, R. (1994) 'Regionalism in the world economy', in Gibb, R. and W. Michalak (ed.), (1994) pp. 1–36.

Gibb, R. and W. Michalak (ed.) (1994) *Continental Trade Blocs*, Chichester: John Wiley.

Greif, A. (1992) 'Institutions and international trade – lessons from the commercial revolution', *American Economic Review*, 82.2, 128–33.

Grossman, G. and K. Rogoff (ed.) (1995) *Handbook of International Economics*, Vol. 3, Amsterdam: North Holland.

Hamilton, C. and J. Whalley (1995) 'Evaluating the impact of the Uruguay round results on developing countries', *The World Economy*, 18.1, 31–50.

Heckscher, E. (1919) 'The effect of foreign trade on the distribution of income', *Economisk Tidskrift*, 21, 497–512, in (trans. and ed.) Flam, H. and M. J. Flanders, *Heckscher–Ohlin Trade Theory*, Cambridge, MA: MIT Press.

Helliwell (1998) *How Much Do National Borders Matter?*, Washington, DC: Brookings Institution Press.

Helpman, E (1998) 'Explaining the structure of foreign trade – where do we stand?', *Review of World Economics*, 134.4, 573–89.

Helpman, E. and P. R. Krugman (1992) *Trade Policy and Market Structure*, Cambridge, MA: MIT Press.

Horvat, B. (1999) *The Theory of International Trade – An Alternative Approach*, London: Macmillan Press.

Jackson, J. H. (1995) The World Trade Organization – watershed innovation or cautious small step forward?', *The World Economy*, 18, 11–31.

Jackson, J. H. (1996) 'Perspectives on regionalism in trade relations', *Law and Policy in International Business*, 27.4, 873–878.

Johnson, H. G. (1965) 'Optimal trade intervention in the presence of domestic distortions', in R. E. Caves *et al.* (eds.), *Trade, Growth, and Balance of Payments*, Amsterdam: North Holland.

Kemp, M. C. (1964) *The Pure Theory of International Trade*, Englewood Cliffs, NJ: Prentice-Hall Inc.

Krugman, P. (1991) *Geography and Trade*, Cambridge, MA: MIT Press and Leuven University Press.

Lawrence, R. Z. (1996) *Regionalism, Multilateralism, and Deeper Integration, Washington*, DC: The Brookings Institution.

Leamer, E. (1984) *Sources of International Comparative Advantage – Theory and Evidence*, Cambridge, MA: MIT Press.

Leamer, E. and J. Levinsohn (1995) 'International trade theory – the evidence', in G. Grossman and K. Rogoff (eds.), (1995), pp. 1339–94.

Lipsey, R. G. (1957) 'The theory of customs unions – trade diversion and welfare', *Economica*, 24, 40–6.

Macbean, A. I. and Snowden, P. N. (1983) *International Institutions in Trade and Finance*, London: Allen & Unwin.

McLaren, J. (1997) 'Size, sunk costs, and Judge Bowker's objection to free trade', *American Economic Review*, 87.3, 400–20.

Meade, J. E. (1955) *The Theory of Customs Unions*, Amsterdam: North Holland.

Ohlin, B. (1924) *The Theory of Trade*, reprinted in H. Flam and M. J. Flanders (ed.), op cit.

Ricardo, D. (1821) *On the Principles of Political Economy and Taxation*, 3rd edn, London: John Murray.

Romer, P. (1994) 'New goods, old theory, and the welfare costs of trade restrictions', *Journal of Development Economics*, 43.5, 5–38.

Sager, M. A. (1997) 'Regional trade arrangements – their role and the economic impact on trade flows', *The World Economy*, 20.2, 239–52.

Samuelson, P. A. (1948) 'International trade and the equalization of factor prices', *Economic Journal*, 58, 163–84.

Smith, A. (1776) *An inquiry into the Nature and Causes of the Wealth of Nations*, vol. 1, ed. by Edwin Cannan, New York: Modern Library, 1937.

Siebert, H. (1987) *Economics of the Environment – Theory and Policy*, Heidelberg: Springer Verlag.

Torstensson, J. (1998) 'Country size and comparative advantage – an empirical study', *Review of World Economics*, 134.4, 590–611.

Trefler, D. (1995) 'The case of the missing trade and other mysteries', *American Economic Review*, 85.5, 1029–46.

UN (1995) *World Economic and Social Survey 1995*, New York: UN.

Viner, J. (1950) *The Customs Union Issue*, New York: Carnegie Endowment for International Peace.

Wan, Jr, H. and N. V. Long (1998) 'Profile: Murray C. Kemp, *Review of International Economics*, 6.4, 698–705.

WTO (1998a) *Trading into Future*, Geneva: WTO Secretariat.

WTO (1998b) *Overview of Developments in the International Trading Environment*, Annual Report by the Director General, WTO Document WT/TPR/OV/4, Geneva: WTO Secretariat.

2
Trade, Environment and Development

2.1 Introduction

The linkages between trade, environment and development remain very significant. With an ever increasing economic globalization and interdependence, these linkages assume much greater importance than in the past. These issues deserve considerably greater focus in the international trade policy regimes. The costs of neglecting these integral dependencies is to engage in misguided trading operations which can precipitate in potentially irreversible environmental and hence economic impediments to sustainability of trade. The role and limitations of trade liberalization are discussed in the next section. The environmental implications of free trade, methods of internalizing environmental costs of trade, and an integration of trade with sustainable development are among the major aspects for discussion in this chapter.

The imperatives of sustainable development, those of international environmental laws, in addition to international trade laws are some of the major issues for discussion in this chapter. Some of the important linkages between trade, environment and development concern those of economic and environmental externalities: the uncompensated effects of trade and other economic activities, with non-local effects and long-term effects. This raises the problems of winners and losers in a broader sense, and also the corresponding management of fair and sustainable trade activities. Non-discrimination and reciprocity in trade relations between countries remain the main pillars of multilateral trade arrangements. It is important to recognize that during the past half century when these concerns led to policies and experiences of global trade governance, much has changed in the area of environment and patterns of development. Economic growth is but one of several conditions for sustainable trade. Trade liberalization with non-discrimination and reciprocity constitute only a set of necessary but not sufficient conditions for comprehensive development and sustainable trade. Accordingly, it is important that the trade mechanisms also are tailored to complement the efforts in the environmental and economic development arena. Only such a framework can ensure patterns of sustainable development wherein sustainable multilateral trade remains an important contributor. This chapter addresses the key issues of trade liberalization, environmental aspects of international trade policies, and provides an integrated framework for sustainable development with the interdependencies incorporating trade and environmental features.

2.2   Trade liberalization and development

In general, trade liberalization leads to greater volume of export–import activities, enhancing economic growth and consumption. Both the production and consumption processes affected by trade liberalization tend to lead to greater pollution on a per capita basis. This is not necessarily undesirable in most economies. However, uncontrolled emissions of pollutants and utilization of non-renewable resources can lead to a series of problems. These problems are typically uncompensated externalities of pollution, like low quality of air at local levels, and greenhouse gas emissions affecting the global problems. In principle, augmented income levels, technological progress and capacities to handle growing environmental problems in this process enable countries and economic entities to cope with offsetting measures. However, the important requirement here is whether such an enhanced potential is indeed utilized to tackle the problems or whether it is simply assumed that these will be addressed. This is the issue in the context of the WTO, which primarily is concerned with promotion of freer non-discriminatory trade, and has so far lagged behind in integrating trade and environmental issues.

Trade liberalization is usually a necessary but not sufficient condition for sustained economic growth, and the latter is usually a

necessary but not a sufficient condition for sustainable development. In other words, trade liberalization is usually preferable
to trade protectionism. In order to ensure sustainable trade and
sustainable development, additional measures are required. These
measures may partly be market-based but additional institutional
interventions may also be relevant. It is the judicious combination
of the market and non-market factors that is expected to lead to the
achievement of multiple objectives of trade liberalization, economic
growth and environmental protection. Pursuit of free trade policies
constitute only a part of the solution but not the whole solution
to these considerations. As pointed out in some of the studies in
Agosin and Tussie (1994), there is inconclusive evidence of the sustainability of economic growth attributable to trade liberalization,
contrary to the claims made in several of the studies of the World
Bank (see, for example, Edwards, 1996). In most of the World Bank
lending programmes to developing countries, trade liberalization is
a usual requirement for their access to credit from this bank. In the
Latin American countries, three-quarters of the conditions comprised trade liberalization related activities during the 1980s and
1990s (World Bank, 1990; Greenaway, 1998).

The fallacy of some the claims about positive correlation between
trade liberalization and economic growth is founded on at least two
aspects: (a) lack of accountability of losses in the environmental
assets in relation to expanded trade, leading to problems of enjoy
now and pay later phenomena; and (b) neglect of second order
effects of trade liberalization. Let us note that most countries have
been parties to the Rio Declaration and later resolutions which seek
to provide for environmental accounting, in addition to the conventional national income accounting, to keep track of unsustainable
losses in the environmental resource assets. In the second feature,
some of these could be positive as in case augmented incomes lead
to enhanced skills and human capital formation, and some are negative as in cases of tropical deforestation and soil erosion (see, for
example, Southgate and Whitaker, 1992). The requirements of complementary activities and institutional settings, development imperatives which seek growth with justice and which adopt optimal
resource and environmental management remain relevant here.

How likely are the trade restrictions or arbitrary tariffs to hurt the
economy or economic welfare? Romer (1994) sought to find a set of

answers. Trade distortion, which forces the trading patterns to deviate from their possible equilibrium under market equalization of demand and supply features, imposes costs on the economy. These inefficiency costs (defined as the difference between the national income without tariff and with tariff), in the case of additional imposition of tariffs, tend to equal (in several economies) the square of the tax or tariff rate.

Trade and economic growth

The fundamental purpose of trade liberalization is to allow price signals to guide production systems for exportables and to use the same to discourage importables; however, trade liberalization in itself will not ensure economic growth (Greenaway, 1998). This is believed to bring the resource allocative system closer to comparative advantage.

The links between openness of trade policies and economic growth for 57 countries during the period 1970 to 1989 was empirically investigated by Wacziarg (1998). The findings suggested that a policy of trade openness has a strong positive effect on economic growth in terms of the implications of accelerated accumulation of physical capital, technological transmissions and macroeconomic policy improvement. Even in those cases where trade enhances consumer welfare and/or domestic producer welfare, it is useful to isolate the relative roles of the features of comparative advantage and the transformation of these into sustainable economic gains from trade for the society as a whole.

In a recent study based on a set of detailed empirical studies of countries during the recent several years, Rodrik (1999) posited the effects of open trading and related economic policies of national economies to assess their workability in favor of the same reference economies: openness by itself is not reliable mechanism to generate sustained economic growth; it will tend to widen income and wealth disparities within countries; and more importantly, it will 'leave countries vulnerable to external shocks'. It is the lack of 'fail safe' mechanisms of accelerated free trade that suggests a reasonable role of managed trade in case of small and underdeveloped economies.

One of the viewpoints linking the benefits of trade liberalization to external debt of a country is to suggest (see Diwan, 1990) that the gains from trade could serve in effect as a collateral on a country's foreign debt. If a country is seeking to increase its creditworthiness, an aggressive export promotion strategy might lead to enhanced foreign exchange earnings and credit/debt rating. The question then is, is such a regime sustainable and what are the externalities and other costs of adopting such an approach. After all, debt is usually not a desirable end product; it can only be means for economic development as long as debt remains within sustainable limits. Economic growth and the environment are related to debt levels.

In a few economies, it has been observed that the positive link between economic growth and the quality of the environment, changes its direction after a peaking in levels of some pollutants like sulphur dioxide. This peak is expressed in terms of the incremental ratio of increase in environmental pollution per unit rise in income. It is useful to recognize that the relationship, wherever it can be established logically and empirically, applies only to a select set of environmental pollutants, and not to the overall environmental quality.

Various methods used in the estimation of the interrelationship between economic growth and environmental quality remain largely too weak in the information base and in analysis to be able to offer any significant policy guidance. The analysis so far suggests that the possibility of economic growth itself taking care of environmental sustainability is extremely remote. This does not imply that economic growth is not a necessary prerequisite for improved quality of life in less developed regions. Economic growth does not automatically lead to a turning point after which increases in economic growth lead to improvements in environmental quality. Similarly, a stationary zero level of economic growth is neither a necessary nor a sufficient condition for SD (Rao, 1999). Also, the role and limitations of export trade-led growth deserve attention in this context.

When export trade-led growth is considered the main engine of economic growth, it is more likely to be unsustainable and is built on a good deal of inherent subsidies of nature, labour and/or other inputs. This is not a form of free trade simply determined by the factors of demand and supply. Rather, it is built on one or more forms of trade distortions and should be critically evaluated. One of the potentially beneficial features of such trade for sustainability could be the usefulness of accelerated economic growth and the relative cost effectiveness of augmenting depleted resources (or substituting the same). The latter is, however, predicated on the assumptions of avoidance of irreversible damages to the environmental resources and lack of regional or global environmental externalities. Thus, it may be useful only in the short-run to emphasize features of export trade-led growth, to tide over budgetary or other resource requirements. A prerequisite for the economic success of such measures is that the economic system is not locked into inflexible capital investments and the substitution effects of such focus on the macroeconomic linkages of development are minimal or do not entail additional transaction costs (as would be the case if, for example, the basic skill formation and human capital aspects are neglected). In the long run, this is unlikely to be sustainable.

It is useful to recall some of the observations of the then WTO Director General Renato Ruggiero at the 1998 WTO Symposium on 'Strengthening Complementarities – Trade, Environment and Sustainable Development':

> Trade liberalization can – and must – be a critical ally of sustainable development. But freer markets alone will not solve all of the complex environmental and social issues we face in today's interdependent world. Freer investment is not a recipe for restoring the stratospheric ozone. Lower tariffs in themselves will not halt the destruction of marine resources.

The next section deals with the interdependencies of international trade and the environment.

2.3   International trade and the environment

It is useful to distinguish three types of environmental externalities which deserve attention in the context of international trade. These are the following (see Runge, 1994, for greater details):

1   local externalities, which have their origin in domestic markets and local production units;
2   transboundary externalities, which have their origin in one domestic market but propagate their impacts to trading or neighbouring regions because of the trading mechanisms; and

3   global externalities, as in the of environmental problems of the
    global commons, to the extent these are attributable to interna-
    tional trade mechanisms.

Trade may be relevant in enhancing environmental quality especially when (see also USOTA, 1992): global environmental externalities are attributable to some of the trade activities, markets are
incapable of incorporating environmental costs in the appropriate
product prices, there exists evidence or a reasonable basis to expect
that trade measure will effect the environmental objectives, and the
environmental benefits exceed the environmental pollution costs. The
problem of missing markets for some of the environmental resources,
and of intergenerational demand items lays foundation for the existence of a wide range of externalities, environmental and economic.

An alternative to reflecting environmental costs in product prices
is the traditional taxation mechanism. When transnational pollution is contributed by production and trade, zero tariffs is usually
not optimal for the economy and the environment. Baumol and
Oates (1988) suggested that there may be a role for tariffs or environmental taxes in order to retain the global economy and its productive resource base at an optimal level on a continuing dynamic
basis. However, the mechanisms of taxation cannot be the global
trade governance issue. These are issues best addressed at the domestic economy level, as long as these do not pose discrimination
problems in multilateral trade systems.

Local and global assimilative features and spread of transboundary
pollution via trade is one of the phenomena that deserves particular
attention. An example in this regard is the trade governing certain
categories of chemicals and pesticides. Some of the developed countries banned these in domestic use but continue to manufacture and
export to other countries. The agricultural products like flowers or
fruits are subsequently exported (after severe local environmental and
human health losses) back to some of the same countries that supplied the domestically prohibited goods (DPGs). So much for global
economic integration without concern for environmental damage to
public health and the environment. Some of the details and the current perspectives on these issues are proposed for Chapters 5 and 7.

Does it help to devise common standards among heterogeneous
trading partners in the name of standardization or harmonization of

environmental standards? Very unlikely, because equalizing environmental standards or their harmonization can interfere with cost-effective methods of production and lead to sub optimal economic solutions to trade and environmental problems. Attempts to equalize abatement costs would conflict with efficient reallocation of pollution-intensive industries toward countries with relatively large environmental endowments (Dean, 1992). Before seeking to harmonize standards, it is essential that DPGs are also banned in international trade. This is a prerequisite for any further attempts to harmonization of environmental standards.

The issue of internalization of these externalities remains a complex one at the policy and implementation levels. The minimum that needs to be done at institutional settings like that of the WTO is to delineate the problems and isolate relative roles of different factors so as to minimize the emerging problems.

One of the comprehensive approaches to the economics of production and trade linkages comes from the concept of meta-production function (Hayami and Ruttan, 1985). This production function is the envelope or collection of usual economic production functions, includes the role of non-marked and intangible ingredients in the generalized production system. If some of the nature's services (including natural resources and waste assimilative capacities of the sinks of the planet Earth) appear free, it does not imply that various economic and other activities can go on forever. The sustainability of the services of nature comes at a cost: cost of replenishment (as in the case of carbon sequestration to control the concentrations of greenhouse gases), or costs of limited access to biodiversity (and its beneficial effects on health and other values, as in the case of extinction of botanical and animal species). The role of the meta-production is such as to encompass the interrelationships of various marketed and non-marketed ingredients in the production of tradable output. Environmental upkeep and internalization of environmental costs of trade is one of the methods of paying attention to these issues. A formal production function and cost minimization approach also tends to provide insights in this regard, as long as the time horizon under consideration is not too short.

Changes in the quality of the environment do affect both the production system as well as consumption system. Pollution itself may alter the composition of goods demanded, for example, because of changing health and productivity implications of local pollution. Trade-based economic growth and/or trade liberalization have implications on the environment. The environmental effect is a combined result of four components (OECD, 1997; Copeland and Taylor, 1994): (a) the scale effect that increases pollution; (b) the composition effect that reflects global specialization in industries and their pollution contribution locally; (c) the technical effect that directs substitution favouring cleaner technologies; and (d) the interaction of these three influences. Trade liberalization combined with appropriate internalization of environmental costs promises the potential to augment global welfare in the short run as well as in the long run, and hence may be sustainable. Scale, technique, and composition (STC) effects of trade liberalization have been advocated in recent trade literature as the main determinants of the environmental effects of trade liberalization. The composition effects refers to the likely changes in the industry composition (greater specialization or other effects of comparative advantage explorations) as a result of trade liberalization, the scale effect is simply the trade expansion implications on the scale of production and trade, and the technique effect is due to changes in the production methods arising out of the above and also because of any other influences of stakeholders. These operate on the system primarily from the production components. The effects via consumption, and the role of endogenous preferences in response to new goods in a given stationary market due to the trade expansion activities, remain significant as well.

Purely from an environmental resource accounting aspect, the export and import activities have differential implications for the Net National Product (NNP). Any of the measures which seek to use NNP as an indicator of sustainability (see Rao, 1999, for details) need to make adjustments toward an imputed income from the depletion of the stock of resources, based on those exported. In case of raw exports of non-renewable resources like oil, this can be constructed from an estimate of the total remaining stock of the resource or resources, and an estimate of the present ratio of the domestic to foreign final consumption of the resources (see also Sefton and Weale, 1996). The relative accuracy of the NNP as a function of this aspect can influence its usability as an indicator of sustainability in economies which have significant export of natural and exhaustible resource based products, and potential changes in the export policies and technology of production with special reference to exports. In a detailed analytical study, Sefton and Weale (1996) established that the implications of international trade are seen as follows. For the resource exporter, the possibility of owning a reserve of exhaustible resource affects the changing rates of prices and extraction with the implication of upward adjustments. The converse holds for the importer, and downward adjustments may be required to the national income in order to compensate for the growing scarcity of the resource. The standard closed-economy results constitute a misapplication of traditional methods. These lead to an under assessment of the income of the natural resource exporting countries and an overstatement of the industrial countries.

International market transmits and enlarges the externalities of the global commons, and policies which ignore these linkages are unlikely to be very useful (Chichilinsky, 1994). Studies which use market analyses focus on the costs of environmental policies tend to ignore uncorrected and/or uncompensated externalities. The details in Smith and Espinosa (1996) clarify how environmental externalities can influence general equilibrium evaluations of trade and environmental policies emphasizes the feedback loop or, more simply, the implications of interactions between agents that arise outside markets because of one or more environmental resources: these interactions can play a direct role in the substitutions that influence outcomes within markets. Since non-marketed environmental resources make independent contributions to production and consumption preferences, they influence the signals for marketed goods from both sides of market transactions.

A recent WTO (1999) study suggests that, to a great extent, trade liberalization is not the primary cause of environmental degradation, nor are trade instruments the first-best policy for addressing environmental problems. A significant part of the relationship between trade liberalization and the environment passes indirectly through effects on levels and patterns of production and consumption. Therefore, the environmental benefits of removing trade restrictions and distortions are also likely to be indirect and not readily identifiable in general terms. However, the study points towards a positive relationship between the removal of trade restrictions and distortions and improved environmental quality, through: (a) more efficient factor-use and consumption patterns through enhanced competition; (b) poverty reduction through trade expansion and encouragement of a sustainable rate of natural resource exploitation; (c) an increase in the availability of environment-related goods and services through market liberalization; and (d) better conditions for international cooperation through a continuing process of multilateral negotiations. While these claims are substantially meaningful, these do not provide any scope for ignoring environmental considerations directly in the trade system to enhance their complementarity and timeliness of actions before any irreversible ecological disturbances are precipitated.

Trade and environment – Uruguay Round effects

The empirical estimates of the implications of the Uruguay Round of trade liberalizations have been explored by Cole *et al.* (1998). The STC effects for several countries and regions on five pollutants nitrogen dioxide (NO_2), sulphur dioxide (SO_2), carbon monoxide (CO), suspended particulate matter (PM), and carbon dioxide (CO_2) were estimated. For nitrogen dioxide and carbon dioxide the combined STC effects are estimated to increase emissions for all regions, except for nitrogen dioxide in the US. For the other pollutants, the STC effects are estimated to lower emissions in most of the developed countries but they increase for the developing countries. Approximations of the monetary costs of increased pollution of the specified five pollutants led to a relatively small percentage of the estimated benefits of the Uruguay Round trade effects: less than 2 per cent for any country. Given the uncertainties associated with the costs, it was suggested that the percentage could also be about three times the above. Since the costs were largely based on rather dated information and are not comprehensive of the public health and related costs of pollution, these tend to be substantially below realistic estimates.

Does trade openness or liberalization enhance environmental quality? There are some answers in rather partial settings. The trade liberalization and consequential environmental effects of the policies of the Uruguay Round of trade negotiations were examined by Cole *et al.* (1998). In most of the developing countries the net increase in each of the pollutants is predicted. In the developed countries the emissions of CO, SO_2, and PM are expected to fall, but those of the remaining two pollutants continue to rise. The policy implications of this assessment are not clear, however. Trade expansion generally contributes to economic growth as well as environmental pollution. The real questions are about the relative cost-effectiveness of income gains and the enhanced potential as well commitment to reduce environmental pollution at source regions and globally. If the importers use the trade liberalization mainly to locate pollution generating production in the exporter countries, the game is not what was intended under the trade negotiations. Appropriate cognizance needs to be taken in future policy reforms in this direction.

What happens to the features of competitiveness if some entities participate in the regulatory or voluntary environmental standard regime and others do not? Free-rider phenomena could arise. This could be particularly valid if the WTO devises trade and environmental rules for its members and the non-members are free to pursue their own policies. The least that can be done is to devise guidelines for the aspiring (about 30) countries seeking WTO membership to provide an environmental as well trade policy regimes in place for eligibility into the WTO process.

Does the lack of environmental regulatory regimes lead to location of polluting industries to the poor countries? The answers depend on the existence of a judicious mix of domestic environmental policies and the ability of trading partners to internalize environmental costs of trade. Polluting industries do not find a place in poorer countries, since every poor country is not equally keen to allow the industry to flourish. Besides, the country-risk assessment is more important for an industrial enterprise to base industry locating decision, than environmental 'safe havens' for dumping pollution. Also, the direct costs of environmental compliance are usually estimated as relatively low: at less than about three per cent of the total costs of operation (Low, 1993). Clearly, a bundle of elements of the total transaction costs is the real determinant affecting investment decisions and environmental costs are but a minor part of such a cost basket.

Deforestation and international trade

Increased incentives for agricultural production and trade tend to contribute to land conversion and deforestation, if the land use policies are not properly defined and enforced. The issue is not to find fault with the incentives or trade of agricultural products, but of land and forest conservation policies and their effectiveness. A number of authors (see e.g. Anderson and Blackhurst, 1992) argue that the design of domestic policies of incentives and disincentives, including the regulation of environmental degradation, is a better arena for policy interventions (affecting exporters and importers) to address potential adverse environmental consequences of trade. This is an oversimplicification of the real problem, and could imply simply shifting of the issue to another arena. The processes affecting international trade include the role of environmental subsidies and trade policies. Unilateral country-level actions to protect environmental resources tend to adversely affect trade and budgets of the poorer and export-dependent economies, as explained later in this chapter. Hence the need for an international policy coordination.

Large-scale exploitation of timber resources poses one of the most significant development-related environmental challenges, but the linkage between this issue and trade is rather complex. The trade–economic growth link for tropical forest production seems rather important in countries like Indonesia and Brazil. Although tropical deforestation is alarming in these and other developing economies, commercial logging of tropical wood is not an instrumental contributor to national growth (less than 1 per cent of Gross Domestic Product (GDP)). Often, deforestation is led both by external markets and by land conversion for agriculture; terms of trade, lack of environmental cost inclusion in export pricing and debt are some of the factors behind the former, whereas these and population pressures are the factors which tend to influence the latter effect. Institutional failures and market imperfections (property right delineation and enforcement, uncertainty and monopolistic practices, and logging contract design and government subsidies) are among the major factors behind the deforestation externalities.

The main argument behind allowing exploitation of resources rather than ban is not to allow artificial depression of domestic prices of commodities or other items covered in the attempted protection. This could lead to loss of revenues and to possible neglect in the maintenance and preservation of the items sought to be preserved. Besides, the enforcement of total can be much more difficult than using a quasi-market approach for harvesting, with the important mandatory provision for reforestation.

Trade liberalization is generally popular in so far as its capacity to expand and accelerate transactions of imports and exports of goods and services is concerned. This mechanism of exchange, in a competitive (ideal) world, should promote efficiency of resource utilization. But the latter can occur only when all the resources are properly accounted for. The traditional economic view of comparative advantage as the basis for exchange of goods and services is a relevant foundation for international trade or its liberalization. But this criterion does not in any way ensure rational and efficient use of ecological resources from the view point of long term sustainability or sustainable development (Rao, 1999). When this does not happen, as is the case whenever there are missing markets for some of the goods and services, trade is likely to be a conduit for adverse environmental externalities. These externalities themselves arise from the phenomena of market failure and institutional failure. Accelerated free trade can multiply the environmental problems, which are usually rooted elsewhere: in the problems of production, consumption, economic inequalities, inefficient pricing of traded goods and services, and the absence of fair terms of trade. The failure to take into account the environmental costs in marketed commodities leads to a divergence between private costs and social costs, and this divergence widens over time and with the scale of operations. International trade can help correct some of the failures through the provision of incentives for environmental protection and promoting efficient use of resources, but it does the opposite in some cases (for details on these issues, see OECD, 1994). The following section provides an approach toward sustainable development. These considerations are relevant in devising appropriate trade policies for the global trading system.

2.4   Sustainable development: a synthesis of linkages

An approach toward integration of international trade and environmental issues is similar to that of an integrated approach toward long-term economic development where the interplay of production and consumption systems with the dynamics of the environment. As long as we recognize these linkages, it becomes more clear

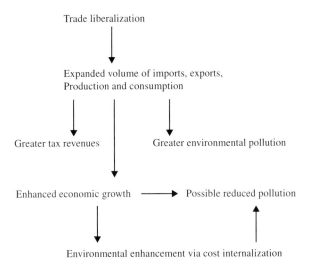

Figure 2.1   Trade and sustainable development

that the feedback mechanisms relate differently to varying methods of trade, production and consumption. Thus, the state of the environment and its feedback effects on the economy are not independent of the elements and mechanisms of trade. Hence the need for a systems approach or synthesis of different major elements. Figure 2.1 indicates an integrated relationship between trade aspects, environment and SD.

Sustainability may be relatively harder to define than indicating features of unsustainability; in fact, the origins of the debate stem from the observations of disturbing trends of possible unsustainability of life on this planet under current trends, with humanity playing a major role against nature. Based on the current knowledge, some of the important symptoms include the following. These are not usually mutually exclusive or independent symptoms, however:

- Greenhouse effect and climate change
- Ozone depletion
- Atmospheric acidification
- Toxic pollution
- Biological species extinction

- Deforestation
- Land degradation and desertification
- Depletion of non-renewable resources like fossil fuels and minerals
- Urban air pollution and solid wastes

Now, can sustainability be defined as a process under which these undesirable features are brought to tolerable levels? This is unlikely, because it is not only these features but also a number of underlying fundamentals of survival that need to be addressed systematically for the purpose of sustainability. A number of alternative definitions of biogeophysical and ecological sustainability have been attempted in literature that provide initial bases to generate the debate and possibly lead to improvements in the clarity of concept policy framework.

The Report of the World Commission on Environment and Development (Brundtland Commission of 1987) contributed to much of the ongoing concern for SD. It stated:

> Sustainable development is development that meets the needs of the present without compromising the ability of future generations to meet their own needs. It contains within it two key concepts: the concept of 'needs', in particular the essential needs of the world's poor, to which overriding priority should be given; and the idea of limitations imposed by the state of technology and social organization on the environment's ability to meet present and future needs (p. 43).

This seems to hold good as a definition, at a general level. Clearly, this approach does address the issue of intragenerational resource distribution, with expressed concern for the poor. However, it is most common that the debate on the issue during the last ten years is very substantially centred only around the intergenerational dimension. Most reports on the theme quote the first sentence, and make no mention of the attendant vital explanation and interpretation. Thus, those writings are possibly less than fair to the spirit of the original contributors. In most studies the debate on sustainable development was interpreted simply in terms of sustainability. The interactions of features of sustainability with those of desirable development patterns were accorded less priority. Within such a framework, the alternatives of sustainability are classified in terms of weak sustainability and strong sustainability, and something in between (like so-called sensible sustainability and other variants; see Serageldin, 1996): Weak Sustainability refers to maintaining total capital intact without regard to the composition of that capital among different kinds of capital; this implies that alternative ingredients of the vector of components of capital are substitutes (within foreseeable ranges of the individual components). Strong Sustainability refers to maintaining every component of the capital vector intact; this assumes that natural and person-made capital are not necessarily substitutes, but are likely complements in the production process.

Weak sustainability is also interpreted as the requirement that the aggregate capital be kept in tact over time, without any decline in the welfare or utility of consumption; strong sustainability is the requirement that natural capital be kept in tact over time. The second version seeks to maintain a critical minimum level of natural capital stocks, specified in physical terms. Weak sustainability permits a large degree of substitutability consistent with the requirements of overall welfare maximization at any given time instant. The substitution is assumed to be such as not to cause an irreversible impediment to the growth processes. It also believes that some of the problems of environmental and ecological degradation may be worth incurring if the benefits of such exploitation render more income for the society than the apparent costs. Thus, the non-monetary losses are sought to be bought off with some monetary levels of potential compensation.

The weak sustainability approaches raise a number of serious objections from environmentalists and ecologists, among others. Implicit in the assumption of high degree of substitution is to treat manufactured capital and ecological capital as close substitutes. Here is what is called the commensurability problem (Rao, 1999): how to trade spotted owl, if need be, with more newsprint? Although this appears a formidable problem, it can be handled with some potential alternatives for decision-making. Some of the issues requiring clarification here include the following: Can we delineate a meaningful region over which this issue needs to be examined, what are the local and regional dimensions of the ecological and economic dependencies, what is a meaningful time frame to resolve the conflict in a cost effective manner, what are the corresponding implications on the stresses and resilience required of the systems and their specific components, who are the current and future winners and gainers?

Many of the differences amongst environmentalists and others who might or might not be as concerned about the environment arise from their adoption of one form or the other of alternative configurations of sustainability. Whether the spotted owl, for example, is some item that can be aggregated into sum total valuation of biological resources and allowed to be substituted by some other owl or some other bird, seems to form a premise for those who advocate Weak Sustainability. This interpretation may not be entirely valid, since even those who advocate this approach admit that there may be certain forms of life or other resource that are irreplaceable and should not be compromised in any aggregation of resources. However, if an aggregation could be allowed this approach forms an intermediary compromise between the two forms of sustainability spanning the continuum of combinations of the two polar or somewhat extreme positions. This is sometimes called the 'sensible sustainability' approach.

SD is the process of socioeconomic development which is built on the sustainability approach (defined above), with an additional requirement that the worth of the capital stocks vector (valued at applicable shadow prices) is maintained constant or undiminished at each time interval for ever (Rao, 1999). A stronger version can be proposed, but this does not arise out of the economic efficiency considerations integrated in the above. Strong Sustainable Development (SSD) may be defined as the process of socioeconomic development which is built on the Strong Sustainability approach with the additional requirement that each individual component of the ecological capital stocks vector is preserved at constant or undiminished levels at each time interval for ever.

A number of pre-specified items are sought to be preserved without any reference to valuations of any type, in either of these two approaches, SD and SSD. This is the feature that bridges the gaps between the traditionally touted concepts of weak sustainability and strong sustainability. This also possesses requisite features to bridge

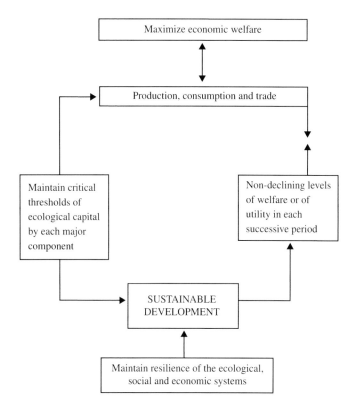

Figure 2.2   Schematic representation of sustainable development

some of the gaps between ecological and economic concepts of sustainability. Figure 2.2 provides an integrated framework toward SD.

The well known concept of sustainability advanced by Robert Solow (1994) stands closer to the SD approach, when it states that a society that invests aggregate resource rents in reproducible capital is preserving its capacity to sustain a constant level of consumption. Solow argued that a concept of sustainability implies a bias toward investment with a general interpretation: just enough investment to maintain the broad stock of capital intact. It does not mean maintain intact the stock of every single thing. Substitution of resources is essential for continued economic progress. This statement does imply the need for continued technical progress, and continued improvements in the resource use efficiencies.

Policy imperatives from the Rio declaration

The 1992 Earth Summit Declaration, also called the Rio Declaration, or the mandate of the World Conference on Environment and Development (WCED), includes a broad set of social, economic and environmental Principles to be followed by the member countries of the UN (thus including the entire set of countries of the WTO membership). Some of these Principles are more closely related to trade issues, as was seen in respect of Principle 16 discussed in the previous section regarding internalization of environmental costs. Among the other important ones for the present context are the following:

Principle 7: 'States shall cooperate in a spirit of global partnership to conserve, protect and restore the health and integrity of the Earth's ecosystem. In view of the different contributions to global environmental degradation, States have common but differentiated responsibilities. The developed countries acknowledge the responsibility that they bear in the international pursuit of sustainable development in view of the pressures their societies place on the global environment and of the technologies and financial resources they command.'

This has implications that the WTO process should participate in some of the multilateral environmental agreements and devise appropriate rules to eliminate perverse incentives to trade in items like DPG. Instead, a fund might be created, as in the case of the Montreal Fund for phase out of aerosols that deplete ozone, to enable developing countries shift to less harmful agrochemicals and related substances.

Principle 12: 'States should cooperate to promote a supportive and open international economic system that would lead to economic growth and sustainable development in all countries, to better address the problems of environmental degradation. Trade policy measures for environmental purposes should not constitute a means of arbitrary and unjustifiable discrimination or a disguised restriction on international trade. Unilateral actions to deal with environmental challenges outside the jurisdiction of the importing country should be avoided. Environmental measures addressing the transboundary or global environmental problems should, as far as possible, be based on international consensus.'

This implies greater coordination requirements between the WTO and the international institutions like the UNEP and others to achieve coherence and economies of scope as well as scale in the relevant activities.

Principle 14: 'States should effectively cooperate to discourage or prevent the relocation and transfer of any activities and substances that may cause severe environmental degradation or are found to harmful to human health.'

Some of the implications of this Principle include not only the short-term concerns like in the Sanitary and Phytosanitary Measures Agreement (see Chapter 4) but also integration of trade and investment measures to incorporate long-term issues like DPGs.

The interface between trade policies under the pursuit of liberalized multilateral trade and the provisions of various multilateral environmental agreements (MEAs) remains the concern of the next few chapters. Chapter 3 maintains focus on MEAs.

2.5   Concluding observations

Trade liberalization merits its importance as long as it does not lead to uncompensated or irreversible environmental degradation. The usual first order effects of trade on incomes and economic growth are not always sustainable. To achieve the later and more specifically, sustainable development, a comprehensive trade–environment–development strategy is called for. Some of the key ingredients are deliberated here, and a few of these are related to the 1992 Rio Declaration and other developments of recent years. It is argued that sustainable trade is a corollary to the phenomena of sustainable development. Complementarity of trade and environment under the multilateral trade regimes cannot be ensured if the protection of the environment is left as resultant indirect contribution of trade mechanisms or when environmental policies are to be pursued only through global environmental agreements. Integrated trade and environmental policies tend to achieve economies of scope and scale, avoid costs of irreversible or expensive degradation of environmental assets and minimize costs of adjustments to ecological disturbances.

References

Agosin, M. R. and D. Tussie (ed.) (1994) *Trade and Growth – New Dilemmas in Trade Policy*, Basingstoke: Macmillan Press.

Anderson, K. and R. Blackhurst (ed.) (1992) *The Greening of World Trade Issues*, New York: Harvester Wheatsheaf.

Baumol, W. J. and W. Oates (1988) *The Theory of Environmental Policy*, Englewood Cliffs, NJ: Prentice-Hall.

Beghin, J., D. Roland-Holst, and D. Van der Mensbrugghe (1994) *A Survey of the Trade and Environment Nexus – Global Dimensions*, OECD Economic Studies #23, 167–92.

Brundtland Commission (World Commission on Environment and Development) (1987) *Our Common Future*, New York: Oxford University Press.

Chichilinsky, G. (1994) 'North–South trade and the global environment', *American Economic Review*, 84, 851–75.

Cole, M. A., A. J. Rayner, and J. M. Bates (1998) 'Trade liberalisation and the environment – the case of the Uruguay round', *The World Economy*, 21.3, 337–47.

Copeland, B. R. and M. S. Taylor (1994) 'North–South trade and environment', *Quarterly Journal of Economics*, 109, 755–87.

Dean, J. M. (1992) 'Trade and the environment – A survey of the literature', in P. Low (ed.), (1992), pp. 15–28: World Bank Discussion Paper #159.

Diwan, I. (1990) 'Linking development, trade, and debt strategies', *Journal of International Economics*, 29.4, 293–310.

Edwards, S. (1996) *The Latin American Debt Crisis*, A publication under the series *The Evolving Role of the World Bank*, Washington, DC: World Bank.

Greenaway, D. (1998) 'Does trade liberalisation promote economic growth?', *Scottish Journal of Political Economy*, 45.5, 491–511.

Hayami, Y. and V. Ruttan (1985) *Agricultural Development – An International Perspective*, Baltimore: Johns Hopkins University Press.

Low, P. (ed.) (1992) *International Trade and the Environment*, Washington, DC: World Bank Discussion Paper #159.

Low, P. (1993) *Trading Free – the GATT and US Trade Policy*, New York: The Twentieth Century Fund.

OECD (1997) *Process and Production Methods*, Report #G97–137, Paris: OECD Secretariat.

Rao, P. K. (1999) *Sustainable Development – Economics and Policy*, Oxford & Boston: Blackwell.

Rodrik, D. (1999) *The New Global Economy and Developing Countries – Making Openness Work*, Washington, DC: Overseas Development Council.

Romer, P. (1994) 'New goods, old theory, and the welfare costs of trade restrictions, *Journal of Development Economics*, 43.5, 5–38.

Runge, C. F. (1994) *Freer Trade, Protected Environment*, New York: Council on Foreign Relations.

Sefton, J. A. and M. R. Weale (1996) 'The net national product and exhaustive resources – the effects of foreign trade', *Journal of Public Economics*, 61.1, 21–47.

Serageldin, I. (1996) *From Concepts to Action – Making Sustainable Development Work*, Washington, DC: World Bank.

Smith, V. K. and J. A. Eispinosa (1996) 'Environmental and trade policies – some methodological lessons', *Environment and Development Economics*, 1.1, 19–40.

Solow, R. M. (1994) 'An almost practical step toward sustainability', in *Assigning Economic Value to Natural Resources*, Washington, DC: National Academy Press.

Southgate, D. and M. Whitaker (1992) 'Promoting resource degradation in Latin America – tropical deforestation, soil erosion, and coastal ecosystem disturbance in Ecuador', *Economic Development and Cultural Change*, 40.4, 787–807.

US Office of Technology Assessment (USOTA) (1992) *Trade and the Environment – Conflicts and Opportunities*, Report #OTA-BP-ITE-94, Washington, DC: US Government Printing Office.

Wacziarg, R. (1998) 'Measuring the Dynamic Gains from Trade', World Bank Working Paper #2001, Washington, DC: World Bank.

World Bank (1990) *Adjustment Lending Policies for Sustainable Growth*, Washington, DC: World Bank.

WTO (1999) *Background Note-Brief History of the Trade and Environment Debate in GATT WTO*, prepared by the WTO Secretariat for March 1999 High Level Symposium on Trade and Environment, Geneva: WTO Secretariat.

3
Multilateral Environmental Agreements

3.1 Introduction

Various MEAs have been worked out from time to time during the past several decades, even before the notable 1972 Stockholm Conference on Human Environment (with its origins on sustainable development). There are currently about 200 MEAs which are addressed to solve the emerging global environmental problems. These have accelerated in their formulation and implementation since about the mid-1980s. In general, MEAs complement some of the international policy measures aimed at the upkeep and protection of environmental resources. Some of the MEAs impose prohibitions on trade, and a few others seek to modulate the methods of production via stipulations of process and product characteristics. This chapter examines some of the issues with particular focus on the interface of trade and environmental measures at the international levels.

The potential conflict between trade and environmental measures, and between developed and developing countries in devising cost-effective measures was recognized at the UNCED in 1992. Accordingly, the Agenda 21(Chapter 2) stated: 'account should be taken of the fact that environmental standards valid for developed countries may have unwarranted social and economic costs in developing countries'. However, Agenda 21 fell short on suggesting any operational guidelines. Trade and environment issues were proposed to be addressed separately in another forum.

Some of the major concerns in this chapter include the role of environmental trade measures in effecting environmental objectives, problems of trade in known harmful products like restricted or banned pesticides, and an assessment of usage of trade measures relative to environmental policies to protect the global environment.

Many of the significant global environmental problems were briefly stated in the context of the issues of sustainability and sustainable development in Chapter 2. A number of regional environmental problems also are of importance in sustaining economic and environmental linkages and hence of sustainable development. This section deals with environmental trade measures to cater to environmental issues, within the framework of the MEAs.

Now let us turn to the role of the environmental trade measures in protecting the environment using trade measures. An environmental trade measure (ETM) may be defined (Charnovitz, 1993) as a restriction of international trade with the explicit stated purpose of promoting an environmental objective. These could be related to processes of production and trade or product characteristics, or both. At least six types of ETMs may be classified. These are based on standards of products/processes, taxes, subsidies, trade restrictions, sanctions, and trade conditionalities. Some apply with *ex-ante* specifications and some others follow as a result of an assessed environmental problem (as in sanctions and conditionalities). A GATT-consistent ETM should possess the features: the measures involved are non-discriminatory, an imported product must be treated no less favorably than the 'like' domestic product, and a product imported from a GATT participating country (GPC) must be treated no less favourably than a like product imported from any other GPC.

Among the regional agreements, NAFTA-related agreements are noteworthy. This was primarily motivated by the need to reduce a wide variety of trade barriers in the region, perhaps more vigorously than under the former GATT regime. A distinguishing feature of NAFTA, relative to most regional trading agreements is its recognition of environmental byproducts and measures to alleviate these problems. The North American Agreement on Environmental Cooperation (NAAEC), and the Border Environment Cooperation Agreement (BECA) were devised after the NAFTA. The NAAEC created two advisory bodies and limited resources for solving emerging environmental problems caused primarily by border trade: Committee on Environmental Cooperation, and Border Environmental

Cooperation Committee (BECC). The BECA requires that BECC adopt procedures to ensure public participation and transparency of project information to enhance stake holder participation in decision-making. The stated principal objective of the BECC is to help 'preserve, protect and enhance the environment of the border region'. However, the limitations of rather insignificant financial resources from the respective governments (for example, only about $56.25 million for each country in 1995–96) and the declining trend of resource allocation led to effective undermining of the intended benefits of some of these agreements and organizations (see also Gantz, 1996).

In general, NAFTA was more considerate of the environment than GATT. The Uruguay Round of Agreements under GATT requires application of necessary cost effective methods for all regulations, whereas under NAFTA these apply only to protection of animals and plants (see also Bryner, 1997). Under NAFTA, parties can establish levels of protection of human health based on 'legitimate' criteria; trade regulations allowed are those that affect 'product characteristics or their related process and product methods'.

Much of the environmental pollution on the Mexican side of the US–Mexico border is the result of the enhanced production and trading activities following NAFTA. Although Mexico has its own stringent environmental laws, their enforcement remains feeble. About 2,100 plants in the region produce goods worth about $20 billion in the late 1990s, and many of these industries are primarily export-oriented 'maquiladora' industries. The maquiladora sector started in 1965 with little environmental regulation and led to an annual increase of about 20 per cent in the number of firms in the border region in Mexico.

At the global level, the limitations of the GATT's approach to environmental concerns became increasingly apparent over the years. To cite a document, the GATT Secretariat's 1991 Report on Trade and Environment declared that 'in principle, it is not possible under GATT's rules to make access to one's own market dependent on the domestic environmental policies or practices of the exporting country'.

A set of important environmental agreements are reviewed in the following section with the objective of identifying their interface with trade policy issues in the global trading regimes.

3.2   Environmental agreements with trade measures

The major international treaties are typically classified into the categories, relating to the planet Earth's emerging problems affecting environmental sustainability and sustainable development. These include: Global Climate Change, Stratospheric Ozone Depletion, Desertification and Land Cover Change, Deforestation, Conservation of Biological Diversity and Ocean-based Bioresources, Transboundary Pollution, and Trade/Industry aspects of the Environment. Almost every one of the agreements comprises provisions involving direct and indirect obligations on the part of the signatories to fulfil the environmental protection features. Some of the international agreements (not necessarily within the purview of the UN system) which provide for trade policies to effect environmental objectives include the following examples: Convention on the Ban of the Import into Africa and the Control of Transboundary Movement and Management of Hazardous Wastes within Africa (1991); Convention on the Regulation of Antarctic Mineral Resource Activities (1988); International Tropical Timber Agreement (1983 and its revisions).

The major Treaties under the UN system which followed the recognition of some of the environmental problems at the 1972 Stockholm Declaration are the following: Convention on Long-Range Transboundary Air Pollution (1979) and its series of seven Protocol Declarations (including two in 1998); Vienna Convention for the Protection of the Ozone Layer (1985) and its follow up with the Montreal Convention of 1987 and the London as well as Copenhagen Amendments of 1990 and 1992, respectively; Basel Convention on the Control of Transboundary Movements of Hazardous Wastes and their Disposal (1989), with Geneva amendment of 1995; Helsinki Convention on the Protection and Use of Transboundary Water Courses and International Lakes (1992); UN Framework Convention on Climate Change (1992) and its Kyoto Protocol of 1997; Convention on Biological Diversity (1992); UN Convention to Combat Desertification in those countries Experiencing Serious Drought and/or Desertification (1994); Lusaka Agreement on Cooperative Enforcement Operations Directed at Illegal Trade in Wild Fauna and Flora (1994); and the Rotterdam Convention on the Prior Informed Consent Procedure for Certain Hazardous Chemicals and Pesticides in International Trade (1998).

The role and effectiveness of the Montreal Protocol has been often cited as a prime example of international cooperation in environmental management, and also in the use of ETMs for achieving environmental objectives (see Benedick, 1998, for a detailed account of the evolution of the Protocol and its measures for implementation). The Protocol set out a time table to effect freeze in developed countries and gradual elimination in other countries of the ozone-depleting substances (ODS); it requires all participating countries to ban exports and imports of these ODS from and to non-participating countries. It also involved creation of a Multilateral Fund to assist developing countries in fulfilling various technological and production tasks to mitigate the depletion of the ozone layer.

A number of international environmental agreements have been in vogue for at least half a century. Any new measures toward trade liberalization need to take into these stipulations, and visualize additional requirements for the sustainability of the environment and economy. Box 3.1 provides a listing of the relevant MEAs. Important aspects of devising international trade policies in consonance with the existing MEAs are: (a) reconciliation of the twin objectives of trade liberalization and promoting environmental protection and (b) achieving complementarity with emerging policies the global commons, especially the concentrations of the greenhouse gases. The first is largely in the arena of local pollution problems while the second is global. Environmental subsidies implicit in some of the exports (especially those of the developing countries) are built upon the assumptions of free environmental goods. These affect domestic environmental features of the exporters, and also the transboundary problems of the environment. In other words, cheap imports and their expansion into the industrial countries may not remain as cheap for anybody in the long run.

In general, MEAs tend to be more effective if either the ETMs envisaged in the MEAs have at least as wide a support as would arise from a similar adoption of trade rules by the WTO. In other words, it is both the number of participating countries and the relative significance of the members in affecting the environment–economy–trade parameters at the global level. It is of importance to note that a few of the MEAs recognize the non-global nature of some of the environmental problems and seek to address these mainly at the national and regional levels. For instance, the role of toxic pesticides is not substantially a global environmental problem but is rather a national problem for the countries which have not phased out these pesticides and other chemicals. But the phase out is better undertaken if appropriate substitutes (like integrated pest management or organic pesticides) or developed and international trade in these chemicals is discouraged. The next section details some of the recent international efforts to address these problems as matter of multilateral cooperation. This is also a standing case of the role of information sharing about the possible hazards of some of the globally traded products. This is also an example of the adoption of differential environmental standards in international trade to enhance trade to the potential detriment of the welfare of the importer. Whereas unilateral ETMs can be restrictive trade practices and WTO-inconsistent, the lack of environmental concern of trade expansion with free play of market factors alone is unlikely to serve the welfare interests of the importers and also of the global environment. Let us recognize that the effects of some of the domestically banned chemicals tend to be internationally transmitted via the trade mechanisms: the hazardous chemicals are exported because of so-called free trade and some of the products (like flowers or other agricultural products) utilizing these inputs and possessing the potential adverse health effects are exported back to the same countries that supplied the chemicals or different sets of countries with no role in the chemical exports. Either way, the transmission of adverse effects is globalized as long as there are trade rules to avoid these problems.

Developing country exporters are paying for the loss of their quality of public health and productivity of their population (both in the short run and in the long run), and importers are expected to incur other costs due to the effects of the loss of the sink capacity of the planet and deterioration in the global commons. The main features of some of these costs are: (a) not entirely monetary; (b) relatively remote in their affliction over time and place; or (c) diffused and stochastic in their sectoral and spatial incidence. None of these features justify that they can be ignored as if they do not matter. When the consequences of environmental neglect do materialize, the costs are direct and extremely high; so will be the costs of effecting any corrective measures. Gradual adjustment of relevant ingredients in the trading policies is less disruptive and cost effective, given the role of the mechanisms of substitution and adaptation. This is feasible when the problems are identified sooner rather than later. Given the current understanding of the issues involved, the WTO/ Committee on Trade and Environment (CTE) will do well to act toward a meaningful international framework for the integration of trade and environmental protection at local and global levels. It is also important to note that the globally accepted soft international law based on the Precautionary Principle should act as the overarching guiding approach in order that any environmental concerns are properly addressed.

Although at least six international commodity agreements (on cocoa, coffee, olive oil and table olives, sugar, tropical timber, wheat) were renegotiated and came into force since the 1992 Earth Summit. These do not attempt to improvise features relevant for sustainable development in a significant way (UN, 1997). However, the International Agreement on Jute and Jute Products contains an objective to give due consideration to environmental aspects by creating the awareness of the beneficial effects of the use of jute as a natural product. The International Agreement on Olive Oil and Table Olives also includes a feature requiring consideration to environmental issues at all stages of the production systems. In the International Tropical Timber Agreement, the objectives of sustainable use of resources, rehabilitation of degraded forest land are emphasized (UN, 1997).

Some of the ETMs are governed by the UN system, and a few others by joint management of institutions and stakeholders. The UNEP-administered Conventions and institutions: the Basel Convention on the Control of Transboundary Movement of Hazardous Wastes; the Convention on Biodiversity; the Convention on International Trade in Endangered Species of Wild Fauna and Flora (CITES); the Vienna Convention; the Montreal Protocol on the Control of Substances that Deplete the Ozone Layer; the Multilateral Fund; the Convention on Migratory Species; the 12 United Nations Environmental Programme (UNEP)-administered Conventions and programmes on regional seas, and UNEP Chemicals, which supports the work of governments towards the recently completed Convention on Prior Informed Consent, as well as supporting the recently initiated negotiations towards a Convention on Persistent Organic Pollutants.

The purpose of CITES is to tackle trade-induced species loss. International trade in wild species is an important cause of species

Box 3.1   Multilateral environmental agreements

Some of the MEAs starting in 1933 were identified by GATT (1992) to contain trade measures. Summarized below are some of the significant Agreements with implications on international trade. Each of these has not been ratified by the same set of countries, however:

International Convention for the Protection of Birds (1950)
International Plant Protection Agreement (1951)
Convention on Conservation of North Pacific Fur Seals (1957)
Convention on Fishing and Conservation of the Living Resources of the High Seas (1958)
Agreement Concerning Cooperation in the Quarantine of Plants and their Protection against Pests and Diseases (1959)
Rio International Convention for the Conservation of Atlantic Tunas (ICCAT) (1966)
Phytosanitary Convention for Africa (1967)
The African Convention on the Conservation of Nature and Natural Resources (1968)
The Benelux Convention on the Hunting and Protection of Birds (1970)
Convention on International Trade in Endangered Species (CITES)of Wild Fauna and Flora (1973)
International Tropical Timber Agreement (1983)
Montreal Protocol on Substances that Deplete the Ozone (1987)
Convention on the Regulation of Antarctic Mineral Resource Activities (1988)
Basel Convention on the Control of Transboundary Movements of Hazardous Wastes and their Disposal (1989)
Convention for the Prohibition of Fishing with Long Drift nets in the South Pacific (Wellington Convention) (1990)
Convention on Biological Diversity (1992)
Lusaka Agreement on Cooperative Enforcement Operations Directed at Illegal Trade in Wild Fauna and Flora (1994)
Agreement for the Implementation of the Provisions of the 1982 UN Convention on the Law of the Sea relating to the

Box 3.1 continued

Conservation and Management of Straddling Fish Stocks and Highly Migratory Fish Stocks (1995)

Kyoto Protocol to the UN Framework Convention on Climate Change (1997)

Aarhus Convention on Access to Information and Public Participation in Environmental Decision Making and Access to Justice in Environmental Matters (1998)

Protocol to the 1979 Convention on Long-Range Transboundary Air Pollution on Persistent Organic Pollutants (1998)

Rotterdam Convention on the Prior Informed Consent Procedure for certain hazardous chemicals and pesticides in International Trade (1998)

Sources: UN (1999), Rao (1999) and GATT (1992)

risk and biodiversity loss. The contributory roles of destruction or land conversion habitats and pollution in biodiversity loss are significant too. The estimated $5 billion annual trade in illegal traffic of wild species is sought to be curbed or minimized with the participation of a number of national governments. Conservative estimates suggest that one in five of all animals are threatened, together with 10 per cent of all birds and plants (UNEP, 1998). Increases in rates of cancer, birth defects, a weakened human-immunity systems and other environmentally related human health effects are linked to the long-term, low-dose exposure to chemicals, pesticides, pollution and other environmental risks.

The Scientific Assessment Panel of the Montreal Protocol reported in 1998 that the 'Montreal Protocol is working'. Because of the Protocol, the level of chlorine and bromine-containing ozone-depleting substances in the lower atmosphere has peaked in 1994, and now has begun to decline. Without the Protocol, ozone-layer depletion in mid-latitudes in the Northern Hemisphere would be at least 50 per cent higher than today, and 70 per cent higher at mid-latitudes in the Southern Hemisphere. Without the Protocol, rates of skin cancer would be higher, while economic and ecological damages to the fisheries, agricultural and other sectors would be several hundreds of billions of dollars as a result of disruption of aquatic food chains and ecological balance.

3.3   Interface with trade policies

The need for coordinating transnational environmental problems is largely contributed both by the existence of physical externalities and by the effects of international trading activities (market externalities). Pollution and market externalities together must constitute the concern of the international trade regimes. A broad global economic integration tends to subsume the realm of environmental integration in such a comprehensive framework. Thus, tariff policies, for example, can be explicitly availed to equalize the effects of trade to bring about the desired optimum levels of economic welfare and environmental protection (some of the illustrative examples of such formulations were advanced by Deardorff, 1997).

Some of the MEAs stipulate trade measures like the ban in specified trade as a method of enforcement of the policies adopted in the agreement. A few others do not impose ban on trade but seek to monitor and resolve compliance issues among the signatories to the agreements. The 1973 CITES and the 1989 Basel Convention specify trade bans of prohibited categories as a measure to enforce environmental and ecological provisions of the agreements. The principle of customary international law, codified in the 1969 Vienna Convention on the Law of Treaties states that in the event of any conflict between trade agreements and environmental agreements, the provisions of later treaty take precedence over those of an earlier treaty. There exists considerable complexity of interpretation when it comes to members of the WTO who are non-signitories of each of the specific MEAs. Article XX(h) of the GATT provides a few guidelines. In general, the Vienna Convention still remains the major source for clarifying most issues involved. Since the WTO is a treaty that has been enacted after several agreements and treaties were entered into (by smaller sets of countries, in most cases) the implications of the WTO Articles are more binding, and the environmental dimensions of these are of particular importance. Since the GATT was not a treaty, provisions of most MEAs took precedence over those of the GATT. Accordingly, it is no surprise that the GATT's recognition and contribution to the issues of environmental protection remained trivial. These details are proposed for a later chapter.

As Howse and Trebilcock (1997) argued, allocative efficiency of the traditional economic approach is but an ingredient of the total system management wherein the economy and the environment are integrally linked and where the efficiency may not entirely be evaluated in terms of market factors. This is due to the relevance and obligations of participants (member countries and trade groups) to comply with various voluntary and mandatory agreements. Thus, some of the trade measures which might restrict free trade need not be welfare decreasing even if they might be viewed as such in a first round effect study ignoring the totality of the system. Trade for its own sake is not necessarily an end product for any society.

The Canadian government position at the March 1999 WTO High Level Symposium on Trade and Environment was that the WTO should develop a set of principles to MEAs that would assist WTO panels in assessing MEA trade measures and international negotiators contemplating the use of trade measures in an MEA. Canada proposed a few qualifying principles for consideration at the WTO Committee on Trade and Environment: (a) the MEA reflects broad-based international support, precisely specified and open to all countries; (b) trade with non-parties to the MEA is permitted on the same basis as parties when non-parties provide environmental protection equivalent to that required by the MEAs; (c) trade measures should not constitute 'arbitrary or unjustifiable discrimination'.

Use of trade measures to protect the environment

It is useful to conclude that, in general, trade measures may be chosen only when: (a) these are cost effective (where the cost includes all elements of transaction costs) relative alternate measures in achieving these environmental objectives; and/or, (b) these act as effective complementary measures to enhance the contribution of other instruments; and/or (c) these are considered indispensable for achieving environmental objectives, even when a set of other feasible alternative measures are brought into operation.

The following section deals with the special problem of trade in chemicals known to be harmful, short term and long term. A set of global measures have been rather recently devised. Much more

additional policy measures are required for a meaningful environmental and trade policy regimes at the global and national levels.

3.4   Harmful chemicals trade

The Global Programme of Action for the Protection of the Marine
Environment from Land-Based Activities, adopted in November 1995
in Washington DC, includes specific provisions to address Persistent
Organic Pollutants (POPs). A regional legally binding Protocol to the
Convention on Long-Range Transboundary Air Pollution (LRTAP)
has been completed under the auspices of the United Nations
Economic Commission for Europe (UN/ECE). It covers 16 POPs (the
12 POPs included in UNEP's first list as well as chlordecone, hexa-
chlorocyclohexane (HCH), hexabromobiphenyl, and polyaromatic
hydrocarbons), and was officially adopted on 24 June 1998 in Aarhus,
Denmark.

A landmark progress in the formulation of a MEA governing the
trade and production of hazardous chemicals, especially those
which are domestically prohibited for use in some of the developed
economies but allowed to export to other countries. The 1998
Rotterdam Convention on Harmful Chemicals and Pesticides (adopted
and signed 11 September 1998) constitutes a major step in moderat-
ing the trade and its effects in cases of some of the listed hazardous
chemicals. Some of these measures are proposed to be availed with
the process of Prior Informed Consent, originally devised about a
decade ago by the Food and Agriculture Organization (FAO), UNEP
and other international organizations.

The Prior Informed Consent (PIC) procedure is a means for formally obtaining and disseminating the decisions of importing countries as to whether they wish to receive future shipments of a certain
chemical and for ensuring compliance to these decisions by exporting countries. The aim is to promote a shared responsibility between
exporting and importing countries in protecting human health and
the environment from the harmful effects of such chemicals.

The current voluntary PIC procedure has been operated by UNEP
and FAO since 1989, based on the amended London Guidelines for
the Exchange of Information on Chemicals in International Trade
and the International Code of Conduct on the Distribution and Use
of Pesticides. The new PIC procedure contained in the Rotterdam

Convention is an improvement of the original procedure and based largely on the experience gained during the implementation of the original. The Convention will enter into force once 50 countries have ratified it. As a first among the MEAs, Governments have agreed to continue to implement the voluntary PIC procedure using the new procedures of the Convention until the Convention formally enters into force. UNEP and FAO have been assigned the responsibility for Secretariat of the Convention.

The Convention requires that hazardous chemicals and pesticides that have been banned or severely restricted in at least two countries shall not be exported unless explicitly agreed by the importing country. It also includes pesticide formulations that are too dangerous to be used by farmers in developing countries. Countries are also obliged to stop national production of those hazardous compounds.

The legally binding treaty will reduce the environmental and health risks posed by hazardous chemicals and pesticides. It will protect millions of farmers, workers, and consumers in developing countries and reduce threats to the environment, according to the Food and Agriculture Organization of the UN (FAO) and the UN Environment Programme (UNEP). This will be achieved by helping governments to prevent chemicals that they cannot safely manage from being imported into their country. If a government does choose to accept an import of a hazardous chemical or pesticide, the exporter will be obliged to provide extensive information on the chemical's potential health and environmental dangers. In this way, the treaty will promote the safe use of chemicals at the national level, particularly in developing countries, and limit the trade in hazardous chemicals and pesticides.

The Rotterdam Convention, a new treaty on trade in hazardous chemicals and pesticides, was signed during a signing ceremony by ministers and representatives from 57 countries and the European Community. The Convention remains open for further signatures for one year. The Convention covers 22 hazardous pesticides. The legally binding treaty will protect the environment and millions of farmers, workers, and consumers from the misuse and accidental release of toxic substances, particularly in developing countries. Many substances that are banned or severely restricted in industrialized countries are still marketed and used in developing countries.

Safer use of chemicals is expected to reduce avoidable or unessential imports of harmful chemicals and pesticides.

In addition to the role of the PIC for select chemicals and their by-products, a set of related chemicals, POPs deserve attention in ETMs and MEAs. The POPs are chemical substances which are persistent, bioaccumulate and pose a risk of causing adverse effects to human health and the environment. It is widely accepted that the use of such persistent, bioaccumulating and toxic substances cannot be considered a sustainable practice. POPs are chemicals or by-products that resist degradation in the environment. They accumulate in the body fat of animals. Concentrations increase for each upward step in the food chain and can reach very high levels in, for example, seals and polar bears. The effects of consuming POPs can be serious, including harmful effects on fertility and embryo development, damage to the nervous system and cancer.

Several toxic and persistent pesticides eventually recognized as POPs were banned by industrialized countries in the 1960s and 1970s after some of their adverse effects had been observed in humans and/or certain animal species. It was thought at the time that such national measures would effectively limit or abolish the problems associated with these chemicals (see also UN, 1991). However, environmental monitoring programmes gradually made it clear that, after an initial decline, concentrations in the environment and in biota were not declining further, according to the UNEP. An internationally binding agreement for implementing global action on select POPs is expected to be ready in early 21st century. The UNEP coordinates further efforts in this direction with the involvement of relevant expertise.

In general, it is maintained that multilateral solutions to transboundary environmental problems, whether regional or global, are preferable to unilateral solutions. Resort to unilateral measures to protect the environment with trade measures runs the risk of arbitrary discrimination and disguised protectionism which could damage the multilateral trading system. Agenda 21 of the 1992 Rio Conference states that measures should be taken to: 'Avoid unilateral action to deal with environmental challenges outside the jurisdiction of the importing country. Environmental measures addressing transborder or global environmental problems should, as far as possible, be based on international consensus.'

Examples of trade measures amongst parties include, the ban of the Basel Convention on the Transboundary Movement of Hazardous Wastes on trade in hazardous wastes, whether for recycling or for final disposal, between essentially OECD and non-OECD parties to the Convention. Examples of trade measures against non-parties include, the trade provisions of the Montreal on Substances that Deplete the Ozone Layer, those of the Basel Convention, and of the Convention on International Trade in Endangered Species, which require parties to apply more restrictive trade provisions against non-parties than to parties.

In discussing the compatibility between the trade provisions contained in MEAs and GATT/WTO rules, the CTE observed that of about 200 MEAs currently in force, only 20 contain trade provisions. No disputes have thus far come to the WTO regarding the trade provisions contained in an MEA. There is a widely held view in the CTE that trade measures agreed to amongst parties to an MEA, even if WTO-inconsistent, could be regarded as *lex specialis* under public international law and ought not to give rise to legal problems in the WTO. Under the principle of *lex specialis*, if all parties to a treaty conclude a more specialized treaty, the provisions of the latter would prevail over those of the former. The issue is that of trade discrimination against non-parties to MEAs.

In terms of institutional developments in recent years, the 1998 Aarhus Convention on Access to Information, Public Participation in Decision-Making and Access to Justice in Environmental Matters provides among other things, institutionalization of reflection of informed public will and the role of the international justice principles via the International Court of Justice. Adoption of this Convention ends a great deal of open-endedness which has been in vogue on several environmental issues at the global level.

3.5   Concluding observations

The role of the MEAs remains at least as much as that of the international trade agreements, whether under the GATT regime or under the current WTO regime. Some of the assertions of the WTO Secretariat suggest that environmental issues are better handled at the MEAs than under WTO trade regimes. This is not always feasible. In general, an integrated approach tends to be cost effective and pragmatic for implementation purpose. The focus of the MEAs has been both on the processes and products affecting the global environment. Some of the trade prohibitions are only partly effective (like some of the exotic bird species), and some are more effective (like the ban and phase-out of the CFCs). It is futile to expect that MEAs alone can help achieve the environmental goals agreed by the participating member countries of the Agreements. It is essential that complementary trade policies and effective measures are devised by the WTO in order to accomplish the trade and environmental objectives in an integrated manner. Also, WTO cannot be a passive spectator in devising trade rules governing, for example, harmful chemicals. Broad-based sustainable development criteria must be taken into account in these processes. If trade policies are simply to maximize trade activities, and environmental agreements are simply to enhance environmental protection and conservation, we are very likely to seek largely conflicting and incongruent policies. The judicious balancing of the policy instruments and their operational implications is to be sought via the common ground of the requirements of sustainable development. This remains the concern for the rest of the book.

References

Benedick, R (1998) *The Ozone Diplomacy*, Cambridge, MA: Harvard University Press.

Bhandari, J. S. and A. O. Sykes (ed.) (1997) *Economic Dimensions in International Law*, New York: Cambridge University Press.

Bryner, G. C. (1997) *From Promises to Performance – Achieving Global Environmental Goals*, New York: W.W. Norton.

Charnovitz, S. (1993) 'A taxonomy of environmental trade measures', *Georgetown International Environmental Law Review*, 6.1, 1–46.

Deardorff, A. V. (1997) 'International conflict and coordination in environmental policies', in Bhandari and Sykes (ed.), (1997), pp. 248–74.

Gantz, D. A. (1996) 'The North American development Bank and the Border Environment Cooperation Commission', *Law and Policy in International Business*, 27.4, 1027–56.

GATT (1992) *Trade and Environment*, Geneva: GATT Secretariat.

Howse, R. and M. J. Trebilcock (1997) 'The free trade – fair trade debate – trade, labor, and the environment', in Bhandari and Sykes (ed.) (1997) pp. 186–234.

Rao, P. K. (1999) *Sustainable Development: Economics and Policy*, Oxford & Boston: Blackwell Publishers.

UN (1991) *Consolidated List of Products Whose Consumption and/or Sale Have Been Banned, Withdrawn, Severely Restricted or Not Approved*, 4th edn, New York: United Nations.

UN (1997) *International Cooperation to Accelerate Sustainable Development in Developing Countries and Related Domestic Policies*, UN Document E/CN.17/1997/12/Add. 1, New York: UN Secretariat.

UN (1999) Information obtained from the UN web site www.un.org (visited 12 March 1999).

UNEP (1989) *London Guidelines for the Exchange of Information on Chemicals in International Trade* (1987, amended 1989), UNEP, Nairobi.

UNEP (1998) *Status of the Convention on Biological Diversity*, Geneva: UNEP Office.

Part II
The World Trade Organization

4
WTO Articles of Agreement and Beyond

4.1 Introduction

The WTO came into existence on 1 January 1995. Its charter is entirely institutional and procedural; it draws upon the GATT and its Uruguay Round resolutions in its Annexes for compliance. The WTO is not a mere replacement or succession of the GATT. The WTO Agreement under the Uruguay Round of GATT Trade negotiations provides a 'common institutional framework for the conduct of trade relations among its members' (Article II of the Agreement). The new institution upgraded several working arrangements and rules with considerable clarity, as in the case of dispute resolution, adoption of resolutions and voiding veto powers for members. Some of the 'birth defects' of the GATT were modified in the WTO. The WTO Charter XVI.1 states clearly that GATT's decisions, procedures and customary practices will be guiding principles to the extent feasible. The WTO budget was estimated (Blackhurst, 1998) at an equivalent of about 10 minutes of the value of the world merchandise trade in mid-1990s.

The WTO remains a rather unique intergovernmental institution governed mainly by a contract between its members. Unlike the GATT, the member countries are not simply 'contracting parties', and the organization is governed by a treaty set up rather than as a 'provisional' organization. The WTO does not describe itself as a 'free trade' institution. Rather, it prefers to be described as a 'system of rules dedicated to open, fair and undistorted competition' (WTO, 1998).

A brief history and salient features of the institutional innovation, implications of some of the provisions of the WTO Agreement, coexistence of the WTO with other international institutions like the World Bank and IMF, the role of environmental concerns within the multilateral trading regimes governed by the WTO are some of the issues for a brief summary in this chapter. Details of the environmental provisions are proposed in Chapter 5. This chapter summarizes some of the major highlights of the WTO mandate, structure, and decision-making methods. Several details of the WTO Agreement are included in Appendix I. The Structure of the WTO is given in Appendix II (Figure A4.1). One of the Agreements covered under the WTO charter, dealing with Technical Barriers to Trade (TBT) is summarized in Appendix III. This TBT lays foundations for admissibility or otherwise of some of the environmental considerations in trade policies at national or multilateral levels. Several additional provisions with implications on the environmental dimension are also seen in various subsidiary agreements under the WTO charter. These are discussed in Chapter 5.

4.2   WTO – an institutional innovation

After about half a century of 'provisional' existence of the GATT, its replacement with the new and more comprehensive treaty-based member-driven institution WTO for the governance of multilateral trading systems is a reflection of the determination of the large number (135 at about the end of the twentieth century) of the countries that are members toward a strengthened global trade regime. The fact that about 30 more countries are waiting to join the WTO is a testimony to the functioning and efficacy of the new institution, shown by the results of the first five years. The charter of the organization and the obligations of the members were formulated in the April 1994 Marrakesh Declaration 'The Results of the Uruguay Round of Multilateral Trade Negotiations'. This comprised 29 legal texts and an additional 28 Declarations/Understandings, leading to a total of 558 pages of statements and stipulations. Very few global agreements tend to be so clearly written and remain focused on the theme and relevant issues. The provision for cooperating with some of the international organizations (including non-governmental organizations) 'that have responsibilities related to those of the WTO' was included in Article V.1 of the WTO Agreement. This is an improvement over the GATT framework which included only the IMF explicitly. Additional details of the working arrangements and their limitations are given in Section 4.3 to follow: 'The defining characteristic of the WTO is that it places more emphasis on contractual rules than on collective action, and it relies on the countries participating in the rules-based trade system to enforce the rules largely on the basis of individual initiative' (Winham, 1998). Thus, the new institution is also called a 'contract' organization. However, the participating countries are members and not simply 'contracting parties' as in the case of the GATT 1947.

The legal structure of the WTO

Whereas the GATT 1947 dealt with trade in goods, the WTO deals with those issues and also the governance of the General Agreement on Trade in Services (GATS), and Trade-Related Intellectual Property Rights (TRIPS), in addition to a well-defined Dispute Resolution Understanding (DSU). The WTO charter includes four Annexes 1 to 4, listed below (see also GATT, 1994 and WTO, 1995). Several additional details regarding these are given in Appendix I:

- Annex 1: Multilateral Trade Agreements. This contains most of the Uruguay Round resolutions and stipulates binding obligations on the members (unlike some or most of the loosely defined obligations under the GATT 1947), detailed below.
- Annex 1A: GATT 1994. This is the revised and comprehensive GATT which includes codes of standards and schedules of tariff concessions. The Agreement on the Application of Sanitary & Phytosanitary Measures deals with some semblance of environmental dimensions.
- Annex 1B: GATS, with its Annexes and specific commitments.
- Annex 1C: TRIPS, with its stipulations.
- Annex 2: Dispute Settlement Procedures.
- Annex 3: TPRM, in existence since 1988 with an emphasis on transparency of trade policies at each individual country level.
- Annex 4: Deals with optional or plurilateral agreements in specific sectors or sub-sectors.

The new institution is unique in several respects, some of which are the following:

1 Specification of contractual obligations such as policy bindings on members.
2 Transparency and coherence in the enforcement of trade policy mechanisms, dispute resolution mechanisms and their compliance.
3 One country–one vote mechanism of governance, providing sufficient voice for the developing countries to advance their viewpoints in trade and economic integration.
4 Time-bound completion of the stipulated methods of compliance of members to various policy agreements and dispute resolution mechanisms.
5 Well-defined time schedules for the governing council or other oversight groups to formulate/implement required activities.
6 TPRM is now a permanent arrangement to integrate trade balance and development aspects of member countries.

WTO articles and the environment

The very first preambular paragraph (see details in Appendix I) in the Agreement for formation of the WTO recognizes the role of sustainable development:

> expanding the production of and trade in goods and services, while allowing for the optimal use of the world's resources in accordance with the objective of sustainable development, seeking both to protect and preserve the environment and to enhance the means for doing so in a manner consistent with their respective needs and concerns.

Despite such proclaimed enthusiasm the rest of the Agreement fails to spell out any details of the role of these critical issues in the governance of the global trade regimes under WTO. This is true of the Annexes and Understandings as well, except in respect of the following two Agreements. The first is on the application of Sanitary and Phytosanitary Measures (SPS) serves at least part of the purpose – but this concerns only a set of immediate public health issues, and does not address the root causes at all. The second is on the Subsidies and Countervailing Measures (SCM), where environmental subsidies of specified nature are exempt from being considered actionable for the purpose of invoking any WTO-inconsistency or attracting a retaliatory action. But this exception, remains a rather feeble concession for environmental consideration to a very marginal degree only. Several additional details of related issues are presented in Chapter 5.

'Benefits of the WTO'?

There are several benefits of the WTO in the governance of the global trading arrangements and consequential economic effects. These benefits are not necessarily all the ones or the only ones listed by the WTO Secretariat (see below). When any meaningful integration of the trade and environmental policies can be accomplished via the WTO regime, it will provide the most significant lasting contribution to the welfare of the global economic and environmental systems to the mutual reinforcement of each other. Such an integration lends stability to the trade regimes, institutions, and lead to maximum net positive benefits of trade.

The WTO Secretariat launched a public information dissemination exercise in 1999. This included display of most of its documents and other information in its electronic web site (www.wto.org) and also a detailed claim of the benefits of the institution to the general public and the world as a whole. One wonders if the stated objectives of the WTO and these ten benefits match in their sense of priorities, identity of objectives, means and ends of the policies and activities. Let us narrate these ten benefits:

1 The system helps to promote peace.
2 The system allows disputes to be handled constructively.
3 A system based on rules rather than power makes life easier for all.
4 Freer trade cuts the costs of living.
5 It gives consumers more choices, and a broader range of qualities to choose from.
6 Trade raises incomes.
7 Trade stimulates economic growth.
8 The basic principles make life more efficient.
9 Governments are shielded from lobbying.
10 The system encourages good government.

The claims of this list do not seem to represent the rationale and mandate of the WTO charter for at least two reasons: (a) peace aspect, lobbying and several other aspects of these claims are not entirely correlated to the charter of the organization; (b) neither sustainable development nor the issue of natural resource and environmental conservation finds a place, despite a clear emphasis on these aspects in the preamble to the WTO Agreement. It is often circulated at some of the WTO deliberations that the environmental matters are better addressed by environmental policies, but does the above listing suggest, for example that matters of peace are better handled via the framework and operations of the WTO? Some of the claims are simply fallacious or only partially meaningful, as in the premise that trade stimulates growth. The question is what mix of trade policies lead to economic growth and on a sustainable basis. Perhaps the WTO Secretariat should focus more on its own charter than attempting a public relations exercise with little validity or credibility on the claims.

The new world body is still to be fully equipped with resources, financial and human. Even after it does get adequately equipped, its interface with a variety of global institutions and stakeholder entities remains an important facet for its efficient functioning. Although the WTO Secretariat itself is supposed to have a minimal role in this 'member-driven' institution, efforts to coordinate global economic activities enable avoidance at cross-purposes. The WTO charter provides for a formal interaction with relevant institutions. The following section deals with the WTO interface with the Bretton Woods institutions, and some of the potential implications.

4.3   The Bretton Woods institutions and the WTO

The WTO signed a formal agreement of cooperation with the IMF and the World Bank in November 1996. The objectives, tasks, and the criteria of performance and their accountability of the international financial institutions, mainly the World Bank and the IMF, are clearly markedly distinct from those of the WTO. Similarly, even the significant differential in the membership (whether viewed in terms of weighted voting power or counted simply in terms of the number of countries) of the two financial institutions relative to that of the WTO makes it reasonable to expect only very limited cooperation between the institutions. The World Bank and IMF

Reports are already considered during the country-specific trade policy reviews. A greater influence of these two institutions on the working methods of the WTO is unlikely to be productive. The IMF and the World Bank should not be allowed to steer the WTO process with backseat driving. Any lopsided attempt to integrate activities or seek to 'harmonize' the same between the different bodies could seriously jeopardize the multilateral trading system contemplated under the WTO, and eventually the new institution itself. The risk of unbounded influence of the two financial institutions is their undue influence over the WTO policies and potential attempts to coerce some of the developing countries with concerted trade policy prescriptions. Such risks can be avoided if the General Council of the WTO adheres to the stipulation and spirit of the Article V.1 whereby the cooperating institutions must be ensured to possess responsibilities similar to those of the WTO. When entering into a cooperation agreement with the IMF and the World Bank, the WTO stated that the agreement was to 'lay the basis for carrying forward the WTO's Ministerial Mandate to achieve greater coherence ... and ... establish new mechanisms by which the institutions can address each other' (WTO, 1996).

Let us consider a plausible configuration. Trade liberalization is usually a condition for an additional loan commitment from the IMF or the World Bank for a borrower country. The objective of the lending institution is to ensure its loan recovery in a stipulated time horizon rather than ensuring a sustainable production and trade pattern of the borrower. The WTO process is not expected to subscribe to these objectives. However, some activities tend to act in a complementary manner, provided the complementarity feature is identified and ensured to be effected (rather than simply assumed to exist) in any cooperation between the institutions. Despite the fact that trade liberalization remains an avowed focus of the operations of the IMF and the World Bank, the 'high conditionality adjustment programmes ... have only rather weak revealed ability to achieve their own objectives' (Killick, 1997).

The OECD, which was established in 1961, in its charter Article 1 states that it is enjoined 'to contribute to the expansion of world trade on a multilateral non-discriminatory basis in accordance with international obligations'. The performance is somewhat close to the desired goals and it remains a plausible cooperator for the WTO.

The WTO Agreement in its entirety is a manifestation of much wider and more comprehensive concerns, relative to the old GATT (now referred to as GATT 1947). The following section deals with some of the salient features of the new system; those relating to the dispute settlement mechanisms and to the environmental considerations, per se, are proposed for Chapter 5.

4.4   Implications of the WTO charter

This section highlights a few major aspects of the articles of agreement for the member countries and the conduct of multilateral trade transactions under the WTO charter. These do not belong to the 'business as usual' GATT 1947 activities and seek greater compliance the WTO regimes. The following selection of four features is far from a comprehensive one, but these tend to highlight the legal and operative structure of the WTO charter.

The first feature listed below enables members to expel a non-complying member from the WTO. The second feature is a vital clarification on the resolution of disputes, especially when these involve the provisions under one or more MEAs, possibly conflicting with the WTO charter. The third relates to the Agreement on Technical Barriers to Trade, which is unique in making a specific mention of the word 'environment' (unlike in the GATT 1947), and also conditions the admissibility of process-based environmental trade measures with its stipulation regarding product-based preferences. The fourth relates to trade policy review mechanism. The mechanism of Trade Policy Reviews was established during the Uruguay Round of Trade negotiations but the objectives need to be examined for possible integration of environmental reviews of trade as well:

1   The Ministerial Conference may decide by a three-fourths majority of the Members that any amendment made effective under this paragraph is of such a nature that any Member which has not accepted it within a period specified by the Ministerial Conference in each case shall be free to withdraw from the WTO or to remain a Member with the consent of the Ministerial Conference (Article IX.5 of the WTO Articles of Agreement).

2   In the event of a conflict between a provision of this Agreement and a provision of any of the Multilateral Trade Agreements, the

provision of this Agreement shall prevail to the extent of the conflict.

3 a The Technical Barriers to Trade Agreement states:

Recognizing that no country should be prevented from taking measures necessary to ensure the quality of its exports, or for the protection of human, animal or plant life or health, of the environment, or for the prevention of deceptive practices, at the levels it considers appropriate, subject to the requirement that they are not applied in a manner which would constitute a means of arbitrary or unjustifiable discrimination between countries where the same conditions prevail or a disguised restriction on international trade, and are otherwise in accordance with the provisions of this Agreement.

b This Agreement also states: 'Wherever appropriate, the standardizing body shall specify standards based on product requirements in terms of performance rather than design or descriptive characteristics.'

4 TPRM: The stated objectives of this are:

to contribute to improved adherence by all Members to rules, disciplines and commitments made under the Multilateral Trade Agreements and, where applicable, the Plurilateral Trade Agreements, and hence to the smoother functioning of the multilateral trading system, by achieving greater transparency in, and understanding of, the trade policies and practices of Members... The assessment carried out under the review mechanism takes place, to the extent relevant, against the background of the wider economic and developmental needs, policies and objectives of the Member concerned, as well as of its external environment. However, the function of the review mechanism is to examine the impact of a Member's trade policies and practices on the multilateral trading system.

The TPRB is established to carry out trade policy reviews. The focus of these discussions shall be on the Member's trade policies and practices, which are the subject of the assessment under the review mechanism; there is no mention of any possible review of environmental aspects of trade measures, or trade aspects of environmental measures resulting from the obligations under MEAs.

Several additional insights can be gained into the provisions and practice of the trade governance rules under the WTO framework, based on the main articles (a few important articles given in Appendix I).

4.5   Concluding observations

Various salient features of the WTO process suggest an increased concern for emerging trends in the global economic interdependence and environmental effects of trade and other economic activities. A number of institutional innovations, partly based on the nearly half a century of the GATT experience, are of significance in enhancing the potential contribution of the WTO as an institution to the multilateral trading system and in fulfilling its more than mercantalist obligations for a sustainable society. The mechanisms for dispute resolution under the new system, reliance on procedures, rules and transparency are important operative ingredients of the system.

Appendix I
Agreement establishing the WTO

The parties to this agreement

Recognizing that their relations in the field of trade and economic endeavor should be conducted with a view to raising standards of living, ensuring full employment and a large and steadily growing volume of real income and effective demand, and expanding the production of and trade in goods and services, while allowing for the optimal use of the world's resources in accordance with the objective of sustainable development, seeking both to protect and preserve the environment and to enhance the means for doing so in a manner consistent with their respective needs and concerns at different levels of economic development,

Recognizing further that there is need for positive efforts designed to ensure that developing countries, and especially the least developed among them, secure a share in the growth in international trade commensurate with the needs of their economic development,

Being desirous of contributing to these objectives by entering into reciprocal and mutually advantageous arrangements directed to the substantial reduction of tariffs and other barriers to trade and to the elimination of discriminatory treatment in international trade relations,

Resolved, therefore, to develop an integrated, more viable and durable multilateral trading system encompassing the General Agreement on Tariffs and Trade, the results of past trade liberalization efforts, and all of the results of the Uruguay Round of Multilateral Trade Negotiations,

Determined to preserve the basic principles and to further the objectives underlying this multilateral trading system, Agree as follows:

Article I: Establishment of the Organization

The World Trade Organization (herein-after referred to as "the WTO") is hereby established.

Article II: Scope of the WTO

1. The WTO shall provide the common institutional framework for the conduct of trade relations among its Members in matters related to the agreements and associated legal instruments included in the Annexes to this Agreement.

2. The agreements and associated legal instruments included in Annexes 1, 2 and 3 (hereinafter referred to as "Multilateral Trade Agreements") are integral parts of this Agreement, binding on all Members.

3. The agreements and associated legal instruments included in Annex 4 (hereinafter referred to as "Plurilateral Trade Agreements") are also part of this Agreement for those Members that have accepted them, and are binding on those Members. The Plurilateral Trade Agreements do not create either obligations or rights for Members that have not accepted them.

4. The General Agreement on Tariffs and Trade 1994 as specified in Annex 1A (hereinafter referred to as 'GATT 1994') is legally distinct from the General Agreement on Tariffs and Trade, dated 30 October 1947, annexed to the Final Act Adopted at the Conclusion of the Second Session of the Preparatory Committee of the United Nations Conference on Trade and Employment, as subsequently rectified, amended or modified (herein-after referred to as 'GATT 1947').

Article III: Functions of the WTO

1. The WTO shall facilitate the implementation, administration and operation, and further the objectives, of this Agreement and of the Multilateral Trade Agreements, and shall also provide the framework for the implementation, administration and operation of the Plurilateral Trade Agreements.

2. The WTO shall provide the forum for negotiations among its Members concerning their multilateral trade relations in matters dealt with under the agreements in the Annexes to this Agreement. The WTO may also provide a forum for further negotiations among its Members concerning their multilateral trade relations, and a framework for the implementation of the results of such negotiations, as may be decided by the Ministerial Conference.

3. The WTO shall administer the Understanding on Rules and Procedures Governing the Settlement of Disputes (hereinafter referred to as the 'Dispute Settlement Understanding' or 'DSU') in Annex 2 to this Agreement.

4. The WTO shall administer the Trade Policy Review Mechanism (hereinafter referred to as the TPRM) provided for in Annex 3 to this Agreement.

5. With a view to achieving greater coherence in global economic policy-making, the WTO shall cooperate, as appropriate, with the International Monetary Fund and with the International Bank for Reconstruction and Development and its affiliated agencies.

Article IV: Structure of the WTO

1. There shall be a Ministerial Conference composed of representatives of all the Members, which shall meet at least once every two years. The Ministerial Conference shall carry out the functions of the WTO and take actions necessary to this effect. The Ministerial Conference shall have the authority to take decisions on all matters under any of the Multilateral Trade Agreements, if so requested by a Member, in accordance with the specific requirements for decision-making in this Agreement and in the relevant Multilateral Trade Agreement.

2. There shall be a General Council composed of representatives of all the Members, which shall meet as appropriate. In the intervals between meetings of the Ministerial Conference, its functions shall be conducted by the General Council. The General Council shall also carry out the functions assigned to it by this Agreement. The General Council shall establish its rules of procedure and approve the rules of procedure for the Committees provided for in paragraph 7.

3. The General Council shall convene as appropriate to discharge the responsibilities of the Dispute Settlement Body provided for in the Dispute Settlement Understanding. The Dispute Settlement Body may have its own chairman and shall establish such rules of procedure as it deems necessary for the fulfillment of those responsibilities.

4. The General Council shall convene as appropriate to discharge the responsibilities of the Trade Policy Review Body provided for in the TPRM. The Trade Policy Review Body may have its own chairman and shall establish such rules of procedure as it deems necessary for the fulfillment of those responsibilities.

5. There shall be a Council for Trade in Goods, a Council for Trade in Services and a Council for Trade-Related Aspects of Intellectual Property Rights (hereinafter referred to as the 'Council for TRIPS'), which shall operate under the general guidance of the General Council. The Council for Trade in Goods shall oversee the functioning of the Multilateral Trade Agreements in Annex 1A. The Council for Trade in Services shall oversee the functioning of the General Agreement on Trade in Services (hereinafter referred to as 'GATS').

The Council for TRIPS shall oversee the functioning of the Agreement on Trade-Related Aspects of Intellectual Property Rights (hereinafter referred to as the 'Agreement on TRIPS'). These Councils shall carry out the functions assigned to them by their respective agreements and by the General Council. They shall establish their respective rules of procedure subject to the approval of the General Council. Membership in these Councils shall be open to representatives of all Members. These Councils shall meet as necessary to carry out their functions.

6. The Council for Trade in Goods, the Council for Trade in Services and the Council for TRIPS shall establish their respective rules of procedure subject to the approval of their respective Councils.

7. The Ministerial Conference shall establish a Committee on Trade and Development, a Committee on Balance-of-Payments Restrictions and a Committee on Budget, Finance and Administration, which shall carry out the functions assigned to them by this Agreement and by the Multilateral Trade Agreements, and any additional functions assigned to them by the General Council, and may establish such additional Committees with such functions as it may deem appropriate. As part of its functions, the Committee on Trade and Development shall periodically review the special provisions in the Multilateral Trade Agreements in favor of the least-developed country Members and report to the General Council for appropriate action. Membership in these Committees shall be open to representatives of all Members.

Article IX: Decision-making

1. The WTO shall continue the practice of decision-making by consensus followed under GATT 1947. Except as otherwise provided, where a decision cannot be arrived at by consensus, the matter at issue shall be decided by voting. At meetings of the Ministerial Conference and the General Council, each Member of the WTO shall have one vote. Where the European Communities exercise their right to vote, they shall have a number of votes equal to the number of their member States which are Members of the WTO. Decisions of the Ministerial Conference and the General Council shall be taken by a majority of the votes cast, unless otherwise provided in this Agreement or in the relevant Multilateral Trade Agreement.

2. The Ministerial Conference and the General Council shall have the exclusive authority to adopt interpretations of this Agreement and of the Multilateral Trade Agreements. In the case of an interpretation of a Multilateral Trade Agreement in Annex 1, they shall exercise their authority on the basis of a recommendation by the Council overseeing the functioning of that Agreement. The decision to adopt an interpretation shall be taken by a three-fourths majority of the Members. This paragraph shall not be used in a manner that would undermine the amendment provisions in Article X.

Amendments to the provisions of this Article and to the provisions of the following Articles shall take effect only upon acceptance by all Members:

Article IX of this Agreement;
Articles I and II of GATT 1994;
Article II:1 of GATS;
Article 4 of the Agreement on TRIPS.

3. Amendments to provisions of this Agreement, or of the Multilateral Trade Agreements in Annexes 1A and 1C, other than those listed in paragraphs 2 and 6, of a nature that would alter the rights and obligations of the Members, shall take effect for the Members that have accepted them upon acceptance by two thirds of the Members and thereafter for each other Member upon acceptance by it. The Ministerial Conference may decide by a three-fourths majority of the Members that any amendment made effective under this paragraph is of such a nature that any Member which has not accepted it within a period specified by the Ministerial Conference in each case shall be free to withdraw from the WTO or to remain a Member with the consent of the Ministerial Conference.

4. Amendments to provisions of this Agreement or of the Multilateral Trade Agreements in Annexes 1A and 1C, other than those listed in paragraphs 2 and 6, of a nature that would not alter the rights and obligations of the Members, shall take effect for all Members upon acceptance by two thirds of the Members.

5. Except as provided in paragraph 2 above, amendments to Parts I, II and III of GATS and the respective annexes shall take effect for the Members that have accepted them upon acceptance by two thirds of the Members and thereafter for each Member upon acceptance by it. The Ministerial Conference may decide by a three-fourths majority of the Members that any amendment made effective under the preceding provision is of such a nature that any Member which has not accepted it within a period specified by the Ministerial Conference in each case shall be free to withdraw from the WTO or to remain a Member with the consent of the Ministerial Conference. Amendments to Parts IV, V and VI of GATS and the respective annexes shall take effect for all Members upon acceptance by two thirds of the Members.

6. Notwithstanding the other provisions of this Article, amendments to the Agreement on TRIPS meeting the requirements of paragraph 2 of Article 71 thereof may be adopted by the Ministerial Conference without further formal acceptance process.

Article XIII: Non-application of multilateral trade agreements between particular members

Except as otherwise provided under this Agreement or the Multilateral Trade Agreements, the WTO shall be guided by the decisions, procedures and customary practices followed by the Contracting Parties to GATT 1947 and the bodies established in the frame-work of GATT 1947.

3. In the event of a conflict between a provision of this Agreement and a provision of any of the Multilateral Trade Agreements, the provision of this Agreement shall prevail to the extent of the conflict.

4. Each Member shall ensure the conformity of its laws, regulations and administrative procedures with its obligations as provided in the annexed Agreements.

6. This Agreement shall be registered in accordance with the provisions of Article 102 of the Charter of the United Nations.

List of Annexes

Annex 1

Annex 1A: Multilateral Agreements on Trade in Goods
General Agreement on Tariffs and Trade 1994
Agreement on Agriculture
Agreement on the Application of Sanitary and Phytosanitary Measures
Agreement on Textiles and Clothing
Agreement on Technical Barriers to Trade
Agreement on Trade-Related Investment Measures
Agreement on Implementation of Article VI of the General Agreement on Tariffs and Trade 1994
Agreement on Implementation of Article VII of the General Agreement on Tariffs and Trade 1994
Agreement on Preshipment Inspection
Agreement on Rules of Origin
Agreement on Import Licensing Procedures
Agreement on Subsidies and Countervailing Measures
Agreement on Safeguards
Annex 1B: General Agreement on Trade in Services and Annexes
Annex 1C: Agreement on Trade-Related Aspects of Intellectual Property Rights

Annex 2: Understanding on Rules and Procedures Governing the Settlement of Disputes
Annex 3: Trade Policy Review Mechanism
Annex 4: Plurilateral Trade Agreements Agreement on Trade in Civil Aircraft
Agreement on Government Procurement
International Dairy Agreement
International Bovine Meat Agreement.

Appendix II: WTO structure

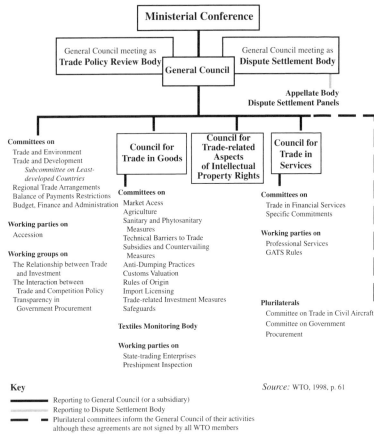

Figure A4.1   WTO structure

All WTO members may participate in all councils etc. except Appellate Body, Dispute Settlement panels, Textiles Monitoring Body, and plurilateral committees and councils.

Appendix III: Agreement on TBT

Recognizing the important contribution that international standards and conformity assessment systems can make in this regard by improving efficiency of production and facilitating the conduct of international trade;

Desiring however to ensure that technical regulations and standards, including packaging, marking and labeling requirements, and procedures for assessment of conformity with technical regulations and standards do not create unnecessary obstacles to international trade;

Recognizing that no country should be prevented from taking measures necessary to ensure the quality of its exports, or for the protection of human, animal or plant life or health, of the environment, or for the prevention of deceptive practices, at the levels it considers appropriate, subject to the requirement that they are not applied in a manner which would constitute a means of arbitrary or unjustifiable discrimination between countries where the same conditions prevail or a disguised restriction on international trade, and are otherwise in accordance with the provisions of this Agreement.

Article 1: General provisions

1.1 General terms for standardization and procedures for assessment of conformity shall normally have the meaning given to them by definitions adopted within the United Nations system and by international standardizing bodies taking into account their context and in the light of the object and purpose of this Agreement.

1.5 The provisions of this Agreement do not apply to sanitary and phytosanitary measures as defined in Annex A of the Agreement on the Application of Sanitary and Phytosanitary Measures.

Article 2: Technical regulations and standards

2.1 Members shall ensure that in respect of technical regulations, products imported from the territory of any Member shall be accorded treatment no less favorable than that accorded to like products of national origin and to like products originating in any other country.

2.2 Members shall ensure that technical regulations are not prepared, adopted or applied with a view to or with the effect of creating unnecessary obstacles to international trade. For this purpose, technical regulations shall not be more trade-restrictive than necessary to fulfill a legitimate objective, taking account of the risks nonfulfillment would create. Such legitimate objectives are, inter alia: national security requirements; the prevention of deceptive practices; protection of human health or safety, animal or plant life or health, or the environment. In assessing such risks, relevant elements of consideration are, inter alia: available scientific and technical information, related processing technology or intended end-uses of products.

2.3 Technical regulations shall not be maintained if the circumstances or objectives giving rise to their adoption no longer exist or if the changed circumstances or objectives can be addressed in a less trade-restrictive manner.

Article 12: Special and differential treatment of developing country members

12.1 Members shall provide differential and more favorable treatment to developing country Members to this Agreement, through the following provisions as well as through the relevant provisions of other Articles of this Agreement.

12.2 Members shall give particular attention to the provisions of this Agreement concerning developing country Members' rights and obligations and shall take into account the special development, financial and trade needs of developing country Members in the implementation of this Agreement, both nationally and in the operation of this Agreement's institutional arrangements.

Article 13: The committee on technical barriers to trade

A Committee on Technical Barriers to Trade is hereby established, and shall be composed of representatives from each of the Members…The Committee shall establish working parties or other bodies as may be appropriate, which shall carry out such responsibilities as may be assigned to them by the Committee in accordance with the relevant provisions of this Agreement.

Article 14: Consultation and dispute settlement

14.1 Consultations and the settlement of disputes with respect to any matter affecting the operation of this Agreement shall take place under the auspices of the Dispute Settlement Body and shall follow, mutatis mutandis, the provisions of Articles XXII and XXIII of GATT 1994, as elaborated and applied by the Dispute Settlement Understanding.

Annex 1: Terms and their definitions

The terms presented in the sixth edition of the ISO/IEC Guide 2: 1991, General Terms and Their Definitions Concerning Standardization and Related Activities, shall, when used in this Agreement, have the same meaning as given in the definitions in the said Guide taking into account that services are excluded from the coverage of this Agreement.

For the purpose of this Agreement, however, the following definitions shall apply:

1. Technical regulation
Document which lays down product characteristics or their related processes and production methods, including the applicable administrative provisions, with which compliance is mandatory. It may also include or deal exclusively

with terminology, symbols, packaging, marking or labeling requirements as they apply to a product, process or production method.

2. Standard
Document approved by a recognized body, that provides, for common and repeated use, rules, guidelines or characteristics for products or related processes and production methods, with which compliance is not mandatory. It may also include or deal exclusively with terminology, symbols, packaging, marking or labeling requirements as they apply to a product, process or production method.

For the purpose of this Agreement standards are defined as voluntary and technical regulations as mandatory documents. Standards prepared by the international standardization community are based on consensus. This Agreement covers also documents that are not based on consensus, and in such cases would be subject to the provisions of this Agreement on regional bodies or conformity assessment systems.

Annex 3

A. For the purposes of this Code the definitions in Annex 1 of this Agreement shall apply.

Substantive Provisions

D. In respect of standards, the standardizing body shall accord treatment to products originating in the territory of any other Member of the WTO no less favorable than that accorded to like products of national origin and to like products originating in any other country.

E. The standardizing body shall ensure that standards are not prepared, adopted or applied with a view to, or with the effect of, creating unnecessary obstacles to international trade.

F. Where international standards exist or their completion is imminent, the standardizing body shall use them, or the relevant parts of them, as a basis for the standards it develops, except where such international standards or relevant parts would be ineffective or inappropriate, for instance, because of an insufficient level of protection or fundamental climatic or geographical factors or fundamental technological problems.

I. Wherever appropriate, the standardizing body shall specify standards based on product requirements in terms of performance rather than design or descriptive characteristics.

References

Blackhurst, R. (1998) 'The Capacity of the WTO to Fulfill Its Mandate', in Krueger (ed.) (1998), op. cit., pp. 31–58.
GATT (1994) *The Final Act of the Uruguay Round*, GATT Focus, 107, Special Issue, May 1994.

Killick, T. (1997) 'Principles, agents, and the failings of conditionality', *Journal of International Development*, 9.4, 483–95.

Krueger, A. O. (ed.) (1998) *The WTO as an International Organization*, Chicago: The University of Chicago Press.

Winham, G. R. (1998) 'The World Trade Organization – institution building in the multilateral trade system', *The World Economy*, 21.3, 349–68.

WTO (1995) *Analytical Index – Guide to GATT Law and Practice*, Geneva: WTO Secretariat.

WTO (1996) *WTO and IMF sign cooperation agreement*, WTO Press Release #62, Geneva: WTO Secretariat.

WTO (1998) *Trading into the Future*, Geneva: WTO Secretariat.

5
Trade Policies and Environmental Provisions

5.1 A brief history

There exists a long history of the attempts to integrate environmental considerations into trade policies at the global level. In more recent years since the 1970s, the 1972 Stockholm Conference on the Human Environment marks an internationally significant event. During the preparatory phase to the Stockholm Conference, the Secretariat of GATT prepared a study entitled 'Industrial Pollution Control and International Trade'. It focused on the implications of environmental protection policies on international trade, reflecting the concern of trade officials at the time that such policies could become obstacles to trade, as well as constitute a new form of protectionism, 'green protectionism'. At the November 1971 meeting of the GATT Council of Representatives, it was agreed that a group on Environmental Measures and International Trade (also known as the EMIT group) be established. However, the group would only convene at the request of Contracting Parties, with participation being open to all. No requests had come forward for its activation during the subsequent two decades. A wake up call in 1991 was motivated by the impending 1992 United Nations Conference on Environment and Development (UNCED), and the need for GATT to contribute in this regard.

During the Tokyo Round of trade negotiations (1973–79), the degree to which environmental measures (in the form of technical regulations and standards) could form obstacles to trade was taken up. However, this Round did precious little for an explicit recognition of complementarity of trade and environmental policies. The Agreement on TBT, earlier known as the 'Standards Code', was negotiated. It called for non-discrimination in the preparation, adoption and application of technical regulations and standards, and for their transparency.

During the Uruguay Round (1986–94), trade-related environmental issues were once again taken up. Modifications were made to the TBT Agreement, and certain environmental issues were addressed in the General Agreement on Trade in Services, the Agreements on Agriculture, SPS, SCM, and TRIPS.

Following the concern expressed by a number of developing countries the fact that products prohibited in developed countries on the grounds of environmental hazards, health or safety reasons, continued to be exported to them. At the 1982 Ministerial Meeting of GATT Contracting Parties, it was decided that the GATT examine the measures needed to bring under control the export of products prohibited domestically (on the grounds of harm to human, animal, plant life or health, or the environment) but which continue to be exported. In 1989, this resulted in the establishment of a Working Group on the Export of Domestically Prohibited Goods and Other Hazardous Substances.

Towards the end of the Uruguay Round a Ministerial Decision on Trade and Environment was adopted (April 1994), calling for the establishment of a CTE. The CTE was to take over from the EMIT group, and from a Preparatory Committee that had been created directly after the adoption of the Ministerial Decision to pave the way for its establishment. A broad-based mandate was agreed upon for the CTE, consisting of identifying the relationship between trade measures and environmental measures in order to promote sustainable development, and of making appropriate recommendations on whether any modifications of the provisions of the multilateral trading system are required. The work programme of the CTE is contained in the Decision and consists of a larger number of issues than those) previously addressed by the EMIT group.

The post-Rio deliberations at the UN Commission on Sustainable Development and the United Nations Conference on Environment and Development (UNCTAD) did not make any significant progress, however. UNEP and UNCTAD declared in 1997 that they work together to provide an integration of trade and environment issues, but these were rather weak players in the international policy games. A relatively more important one was the institution of the GATT, which had its final act of the Uruguay Round of Trade Negotiations (these negotiations lasted about seven years and lead to the Marrakesh Declaration). The GATT had very limited competence on environmental issues. Besides, in the frenzy to conclude the negotiations after seven long years of deliberations, GATT ministers left out the long-term mechanisms for the WTO to integrate trade and environmental issues. The WTO CTE is currently the focal point for the integration of trade and environmental issues. The terms of reference of this CTE are detailed in Appendix I, and the some of the highlights of its conclusions are discussed in Section 5.3. Before examining further the CTE work, it is useful to summarize various important provisions under GATT/WTO framework for environment in the context of multilateral trade. These aspects are discussed in the next section.

5.2   GATT/WTO environmental provisions

The original GATT 1947 did not mention the word 'environment' anywhere in the text. However, a number of GATT articles are of direct relevance to trade-related environmental issues. These include GATT Article I and III on non-discrimination, as well as certain sections of GATT Article XX on General Exceptions, discussed in chapter 2. Under WTO Agreements, environmental concerns are addressed in a number of different Agreements. For instance, the Agreement on Subsidies and Countervailing measures treats as non-actionable any subsidy or government assistance to industry amounting up to 20 per cent of the cost of adapting existing assets and services to new environmental regulations. Several other provisions in different Agreement under the WTO purview are summarized below. Box 5.1 provides a summary of the significant environmental measures provided for dealing with environmental factors in various agreements.

   The Uruguay Round leading up to the WTO did add the issue of Harmonization as a desirable mechanism for reducing arbitrary interpretation of trading standards and their codification. Let it be clear that this 'harmonization' has nothing to do with required harmonization of trade and environmental policies. The WTO framework only seeks to ensure that domestic business regulations of a member country do not exceed international standards, based on a codification of the later. This remains a complex issue in terms of its objectives, means of achieving the same and their environ-mental implications. These aspects will be addressed in Chapter 7.

The provisions under the GATT under Article XX (b) allow a WTO member country to place its public health and safety and national environmental goals ahead of its general obligation not to enhance trade restrictions or utilize discriminatory trade measures. These measures are allowed only to the extent they are 'necessary to pro-tect human, animal or plant life or health'. The original Article came into existence in the late 1960s when due recognition of the environmental factors was not a dominant social or economic para-digm. Accordingly, it seems, there was no explicit mention of the word 'environment'. Relevant modifications in Article XX (b) are

---

Box 5.1   Summary of environmental provisions

Some of the main provisions in the WTO agreements dealing with environmental issues

- GATT Article XX: policies affecting trade in goods for protect-ing human, animal or plant life or health are exempt from normal GATT disciplines under certain conditions.
- Technical Barriers to Trade (i.e. product and industrial stan-dards), and Sanitary and Phytosanitary Measures (animal and plant health and hygiene): recognition of some of the envi-ronmental objectives.
- Agriculture: environmental programmes exempt from cuts in subsidies.
- Subsidies and Countervail: allows one-time subsidies, up to 20 per cent of firms' costs, for adopting to new environmental laws.
- Intellectual property: governments can refuse to issue patents that threaten human, animal or plant life or health, or risk serious damage to the the environment (TRIPS Article 27).
- GATS Article 14: policies affecting trade in service for protect-ing human, animal or plant life or health are exempt from normal GATS disciplines under certain conditions.

essential in light of modern knowledge (see also Rao, 1999; Schoenbaum, 1997).

Another important provision is Article XX (g), details given below. This allows WTO members to take action to conserve exhaustible natural resources. Based on a Tuna/Dolphin dispute between Mexico and the US, it was clarified that the Article does have extra territorial effect but not extra jurisdictional effect. The Tuna/Dolphin II Panel of the GATT endorsed national measures designed to protect extra territorial resources (see Chapter 6, and also Cheyne, 1995). Several of the issues in the use of trade measures to protect the environment and the decisions of the dispute panels at the GATT/WTO levels are proposed in Chapter 6. For the present, the potential application of some of the integrative principles remains important.

In the preamble to the Marrakesh Agreement Establishing the World Trade Organization, reference was made to the importance of working towards sustainable development. It states that WTO Members recognize:

> that their relations in the field of trade and economic endeavor should be conducted with a view to raising standards of living, ensuring full employment and a large and steadily growing volume of real income and effective demand, and expanding the production of and trade in goods and services, while allowing for the optimal use of the world's resources in accordance with the objective of sustainable development, seeking both to protect and preserve the environment and to enhance the means for doing so in a manner consistent with their respective needs and concerns at different levels of economic development.

The following is summary of various principles and provisions under the WTO charter to address environmental issues within the context of multilateral trade; for more details see the WTO document WTO (1999).

The criterion of non-discrimination is the main principle on which the rules of the multilateral trading system are founded. With respect to trade-related environmental issues, the principle ensures that national environmental protection policies are not adopted with a view to arbitrarily discriminating between foreign and domestically produced like products, or between like products imported from different trading partners. Therefore, it prevents the abuse of environmental policies, and of their usage as disguised restrictions on international trade.

Article XX on general exceptions of GATT

'Subject to the requirement that such measures are not applied in a manner which would constitute a means of arbitrary or unjustifiable discrimination between countries where the same conditions prevail, or a disguised restriction on international trade nothing in this Agreement shall be construed to prevent the adoption or enforcement by any contracting party of measures...

(b) necessary to protect human, animal or plant life or health...

(g) relating to the conservation of exhaustible natural resources if such measures are made effective in conjunction with restrictions on domestic production or consumption.'

The chapeau of Article XX is designed to ensure that the GATT-inconsistent measures do not (a) result in arbitrary or unjustifiable discrimination, and/or (b) constitute disguised restriction on international trade. The provision of the article as well as its interpretation when trade disputes arose assumed importance over the years. It is regrettable that this Article XX still has no provision for the term 'environment' for its protection, and it refers to 'exhaustible resources' rather than non-renewable resources. No doubt, the latter was not in common use at the time of the GATT 1947, but it is more than overdue to revise the terminology so that we do not only concern ourselves with 'exhaustible resources' but also pay adequate attention to non-renewable resources in the interests of the environmental, ecological and ecosystem services.

Article XIV on general exceptions of GATS

Negotiated during the Uruguay Round, GATS contains a General Exceptions clause in Article XIV, similar to that contained in GATT Article XX. The Article starts with a chapeau that is identical to that of GATT Article XX. In addressing environmental concerns, GATS Article XIV (b) allows WTO Members to adopt GATS-inconsistent policy measures if this is 'necessary to protect human, animal or plant life or health' (and is identical to GATT Article XX (b)). However, this must not result in arbitrary or unjustifiable discrimination and must not constitute disguised restriction on international trade. Again, this Agreement which came into existence after four decades of the GATT 1947 experience felt shy of using the term 'environment' for protection purposes.

The Agreements on TBT

The Uruguay Round Agreement on TBT, seeks to ensure that technical regulations and standards, as well testing and certification procedures, do not create unnecessary obstacles to trade. The Agreement recognizes the rights of countries to adopt such measures, to the extent they consider appropriate – for example, for the protection of human, animal or plant life or health, or for the protection of the environment. Moreover, Members are allowed to take measures to ensure that their standards of protection are met, known as conformity assessment procedures. Non-discrimination in the preparation, adoption and application of technical regulations, standards, and conformity assessment procedures is one of the main principles of the Agreement. The requirement of transparency of these measures, through their notification to the WTO Secretariat and the establishment of national enquiry points, is an important feature of the Agreement.

The Agreement on SPS

Sanitary measures deal with food safety and animal health aspects. Phytosanitary measures deal with plant life issues. The Uruguay Round Agreement on SPS, addresses the application of food safety, animal and plant health regulations. It recognizes Member's rights to adopt SPS measures but stipulates that they must be based on science, should be applied only to the extent necessary to protect human, animal or plant life or health, and should not arbitrarily or unjustifiably discriminate between members where similar conditions prevail. The Agreement complements the TBT Agreement. This Agreement deals essentially with short-term direct health-related issues only. An extract of the SPS Agreement is provided in Appendix II.

The Agreement seeks to ensure that the health and safety standards are based on scientific evidence, and possible harmonization of standards based upon the recommendations of international scientific organizations such as the Codex Alimentarius Commission and the International Plant Protection Organization. It is useful to recognize the following, as pointed by Ingersent *et al.* (1995):

> it is virtually inevitable that equivalent imports will be required to meet the same standards and that overseas suppliers who are unable to do so will be 'discriminated against', possibly in the face of different and less exacting scientifically based international standards...the Agreement lacks the teeth needed to enforce adherence to international standards where these are in conflict with national preferences.

The Agreement on TRIPS

The Uruguay Round TRIPS Agreement makes explicit reference to the environment in Section 5 on Patents. Article 27 (2) and (3) of Section 5 state that Members may exclude from patentability inventions, whose prevention within their territory is necessary to protect, among other objectives, human, animal or plant life or health or to avoid serious prejudice to the environment. Under the Agreement, Members must provide for the protection of plant varieties either by patents or by an effective *sui generis* system or by a combination of the two.

These provisions are designed to address the environmental concerns related to the protection of intellectual property. The Agreement allows Members to refuse the patenting of inventions which may endanger the environment (provided their commercial exploitation is prohibited as a necessary condition for the protection of the environment), as well as to exclude from patentability plants or animals. Members must provide for the protection of different plant varieties, for the purposes of biodiversity, through patents or other effective means referred to in the Agreement.

The Agreement on SCM

The WTO Committee on SCM drafted the Agreement on its title theme in 1998. This specified certain subsidies to meet new environmental requirements and also research and development activities as 'non-actionable' or disputable by other WTO members; some of these details are furnished later in this section. The Agreement on Subsidies, which applies to non-agricultural products, is a Uruguay Round Agreement designed to regulate the use of subsidies. Under the Agreement, certain subsidies are referred to as 'non-actionable'.

These are generally permitted by the Agreement. Under Article 8 of the Agreement on non-actionable subsidies, direct reference is made to the environment. Amongst the non-actionable subsidies mentioned, are subsidies to promote the adaptation of existing facilities to new environmental requirements imposed by law and/or regulations which result in greater constraints and financial burden on firms (Article 8(c)). Such subsidies, however, must meet certain conditions. Making such subsidies non-actionable enables Members to capture positive environmental externalities when they arise, and minimize negative externalities. Some of the non-actionable subsidies such as those on Research and Development (R&D) are limited to one time allowance of 20 per cent of costs of adaptation of facilities and equipment to new environmental laws. This is not an optimal provision in so far supporting environmental innovation is concerned. It is essential to provide a more detailed set of clearly delineated subsidy incentives in the 'non-actionable' category of economic activities on a continuing basis to both process and product R&D.

The Agreement on Agriculture

Adopted during the Uruguay Round, the Agreement on Agriculture seeks to reform trade in agricultural products and provides the basis for market-oriented policies. In its Preamble, the Agreement reiterates the commitment of Members to reform agriculture in manner which protects the environment. Under the Agreement, domestic support measures with minimal impact on trade (known to as 'green box' policies) are excluded from reduction commitments (contained in Annex 2 of the Agreement).

The post-WTO launching era has its focus on the WTO Committee on various potential developments to integrate trade and environmental issues in a largely complementary arena, besides coming up with pragmatic measures to devise multilateral trade policies that minimize damages to the quality of the environment (in both stock and flow terms). The next section deals with the work of the CTE and the need for further progress in the desired directions.

5.3   CTE

The key entity under the WTO is the CTE. The WTO Committee on Trade and Environment was constituted in 1995 to address the following aspects, among others initially: (a) the relationship between the provisions of the multilateral trading system and trade measures for environmental purposes, including those pursuant to multilateral environmental agreements; (b) relationship between environmental policies relevant to trade and environmental measures with significant trade effects and the provisions of the multilateral system, including charges and taxes, and requirements relating to products, packaging, labelling and recycling for environmental purpose; and (c) the effects of environmental measures on market access, especially for developing countries. Much of the desired work is still in progress as of 1999.

The Marrakesh Declaration called for the CTE to examine the role of the WTO in relation to the links between environmental measures and new trade agreements reached in the Uruguay Round negotiations on services and intellectual property. The Ministerial Decision to establish CTE in 1994 stated that the CTE was proposed for establishment:

> Considering that there should not be, nor need be, any policy contradiction between upholding and safeguarding an open, non-discriminatory and equitable multilateral trading system on the one hand, and acting for the protection of the environment, and the promotion of sustainable development on the other,
>
> Desiring to coordinate the policies in the field of trade and environment, and this without exceeding the competence of the multilateral trading system, which is limited to trade policies and those trade-related aspects of environmental policies which may result in significant trade effects for its members,
>
> Recalling the preamble of the Agreement establishing the World Trade Organization (WTO), which states... Noting: the Rio Declaration on Environment and Development, Agenda 21, and its follow-up in GATT, as reflected in the statement of the Chairman of the Council of Representatives to the Contracting Parties at their 48th Session in December 1992.

Since its establishment in the beginning of 1995, the CTE has discussed all of the items contained in its work programme. It reached a few conclusions which it presented to the Singapore Ministerial Conference in December of 1996. The following is a presentation of the main conclusions reached, in addition to the current state of debate on the various items of the CTE's mandate (WTO, 1999).

The CTE is composed of all WTO Members and a number of observers from inter-governmental organizations, and reports to the WTO's General Council. It first convened in early 1995 to examine the different items contained in its mandate. In preparation for the Singapore Ministerial Conference, which took place in December 1996, the CTE summarized the discussions which it held since its establishment as well as the conclusions which it reached in a report presented at the conference. Some of these 'conclusions' are summarized below:

1 WTO is not an environmental protection agency, and that its competence for policy coordination in this area is limited to trade policies, and those trade-related aspects of environmental policies which may result in a significant effect on trade.
2 GATT/WTO Agreements already provide significant scope for national environmental protection policies, provided that they are non-discriminatory; secure market access opportunities are essential to help developing countries work towards sustainable development.
3 Multilateral cooperation in the form of MEAs constitutes the best approach for resolving transboundary (regional and global) environmental concerns.

It was also suggested in a recent document WTO (1999) that multilateral environmental solutions to global environmental problems reduce the risks of arbitrary discrimination and disguised protectionism, and reflect the international community's common concern and responsibility for global resources. In terms of the recent contributions of the CTE, the WTO Secretariat (WTO website www.wto.org, January 1999) would like to assert that the CTE has 'broken a new ground', on these issues. And this ground consists of the following two observations: (a) the relevance of any additional measures was not considered necessary (other than the existing Article XIV (b)), according to the first report of the CTE; and (b) it was suggested that there is need to examine more on the relationship between the TRIPs and the Convention on Biological Diversity (CBD). In its first report submitted at the 1996 WTO Ministerial Conference in

Singapore, the CTE feebly expressed the need to integrate the two title words of the Committee.

Discussions in WTO committee on trade and environment (WTO, 1999)

Trade-related environmental policies: subsidies

Subsidies have the potential to contribute either positively or negatively to the environment. They may contribute positively when they capture positive environmental externalities. On the other hand, they may contribute negatively if they cause environmental stress (by, for instance, encouraging the overuse of certain natural resources). In the areas of agriculture and energy, subsidies are widely viewed as being trade distorting, and as, in some instances, being the cause of environmental degradation. Environmentalists have suggested that multilateral trade rules should incorporate greater flexibility for providing subsidies to encourage activities or technologies which have a beneficial impact on the environment.

During the Uruguay Round both the positive and negative contributions which subsidies may make to the environment were considered, and a number of new disciplines, as well as exemptions, were included in the Agreements on Agriculture and Subsidies and Countervailing Measures. Both these Agreements make certain exemptions for environmental subsidies. Under the Agreement on Agriculture, environmental subsidies may be exempt from domestic support reductions when certain conditions are met, and under the Agreement on Subsidies they may be exempt from countervailing duties as well as panel disputes provided that certain conditions are also met.

The revised rules for export subsidies provided in the Agreement on Subsidies, whereby taxes on energy used to produce exports can be refunded without such refunds being treated as an export subsidy.

The environmental review of trade agreements

The US and Canada have prepared reviews of the North American Free Trade Agreement and of the Uruguay Round. The US recommended

the use of environmental reviews of trade agreements by governments at the national level. Further analysis of the use of these
reviews will need to be undertaken in future CTE discussions for the
benefit of all members.

Eco-labelling has been one of the most controversial aspects of the
CTE's work and has received considerable attention. The CTE recognized that well-designed eco-labelling programmes can be effective
environmental policy instruments which may be used to foster
environmental awareness among consumers. However, it noted that
such schemes raise significant concerns regarding their possible
trade effects.

While some are based on a single criterion, others are based on
life-cycle analysis, i.e. the consideration of the environmental effects
of products starting from their process of production until their
final disposal. In practice life-cycle analysis is not easy to conduct,
and labels following the latter approach are frequently based on criteria that relate to only a few aspects of a process of production or of
a product. This creates the potential for trade restriction.

While the TBT Agreement applies to labelling requirements, it
exerts stronger control on mandatory labels (referred to as 'technical
regulations' under the Agreement) then it does on voluntary ones
(referred to as 'standards'). Furthermore, the extent to which the
Agreement applies to process-based labels (instead of to the characteristics of the products themselves) is unclear and has been the subject of some discussion in the both the CTE as well as the
Committee on Technical Barriers to Trade, which administers the
TBT Agreement. Currently, a major challenge to the effectiveness of
the TBT Agreement is the increasing use (not only in the area of the
environment) of process-based, as opposed to product-based, regulations and standards. The CTE has supported the view that environmental concerns vary across countries, and that the eco-labels
developed by different countries need not be based on the same criteria. In the context of international trade, this raises the issue of
the comparability of different criteria and of conformity assessment
procedures.

Environmental charges and taxes are increasingly being used by
WTO Member governments for the pursuit of national environmental policy objectives and for 'internalizing' domestic environmental

costs. WTO rules discipline the way in which governments impose internal taxes and charges on traded goods, when either imposed on imported products or rebated on exports. This is an issue of considerable interest and importance to trade and environment policy-makers in the context of proposals to increase taxes on environmentally sensitive inputs to production, such as energy (i.e. carbon taxes) and transportation.

The work programme in the 'Decision on Trade in Services and the Environment' notes that: 'since measures necessary to protect the environment typically have as their objective the protection of human, animal or plant life or health, it is not clear that there is a need to provide for more than is contained in paragraph (b) of Article XIV'. In order to determine whether any modification of GATS Article XIV is required to take account of such measures, the Decision requested the CTE to examine and report, with recommendations if any, on the relationship between services trade and the environment including the issue of sustainable development. It also asked the CTE to examine the relevance of inter-governmental agreements on the environment and their relationship to the GATS.

Discussions in the CTE on the GATS has not led to the identification of any measures that Members feel may need to be applied for environmental purposes to services trade which would not be covered adequately by GATS provisions, in particular Article XIV(b).

The saga of DPG and the need for a multilateral trade policy regimes fell on deaf ears, as far as the CTE is concerned; see Box 5.2 for detailed account. One cannot but notice the 'bureaucratease' of the expressions on a major environmental and public health issue. It is doubtful if there is an appreciation of the adverse implications both for the importers and for the exporters at the respective country levels. On the set of environmental trade measures and related issues, the CTE recommended that trade-related environmental measures notified to the WTO under different agreements, be compiled in a single database for easier access and for the greater transparency of this particular category of measures. A WTO Environmental Database has been established and is periodically updated by the WTO Secretariat.

Box 5.2   DPG and CTE – much ado about nothing?

The GATT has examined the issue of the export of DPG as early as 1982. Concern was raised by a number of developing country Parties to GATT about the fact that goods were being exported to them, when their domestic sale in exporting countries had been either prohibited or severely restricted on health and environmental grounds. This raised ethical and health concerns. From the point of view of these countries, these needed to be addressed within the bounds of the multilateral trading system. In the 1982 Ministerial Meeting of GATT Contracting Parties, it was agreed that the GATT should examine the issue, and that all Parties should begin to notify the GATT of any goods produced and exported by them but banned for health reasons by their national authorities for sale in their domestic markets. While 2the notification system began to function following this Decision, Parties tended to notify DPG whose export had also been prohibited rather than the ones which they continued to export. The notification system was ignored, and no notifications were received after 1990, despite the fact that the 1982 Decision remains in force.

In 1989, a Working Group on the Export of DPG was established in GATT. The Working Group met 15 times between 1989 and 1991, when its mandate expired, but failed to resolve the issue. In the 1994 Ministerial Decision on Trade and Environment it was agreed to incorporate DPG into the terms of reference of the CTE.

It may be noted that a few international instruments under MEAs address the export of DPG, such as the Basel Convention on the Transboundary Movement of Hazardous Wastes. However, the issue of chemicals that are of 'severely restricted' category or other potentially dangerous agrochemicals are not part of the list under these MEAs.

Collectively, the CTE has stated that while there is a need to concentrate on the role which the WTO can play on this issue, it is important neither to duplicate nor to deflect attention from the work of other specialized inter-governmental organizations: 'It also recognized the important role that technical assistance and transfer of technology related to DPG whose trade is allowed, can play

Box 5.2 continued

in both tackling environmental problems at their source and in helping avoid unnecessary additional trade restrictions on the products involved.' It stated that WTO Members should be encouraged to provide technical assistance to other Members, especially developing and least-developed countries, either bilaterally or through inter-governmental organizations. This would assist these countries in strengthening their technical capacity to monitor and, where necessary, control the import of DPG.

Based on a Secretariat note prepared on the information already available in the WTO on the export of DPG, some delegations requested that the DPG notification system that had been in existence between 1982 and 1990 be revived, particularly as the Decisions taken to establish it remain in force today. However, no decision on this issue was taken in the CTE. Thus, the CTE has been of little use in devising a policy guidance on a problem of global importance.

Source: WTO (1999)

5.4 National environmental laws and international trade agreements

There are number of national laws governing environment or resource conservation in many of the developed and developing countries. Some of these, in their application and interpretation, tend to conflict with the dated provisions of the GATT Article XX which did not include the word 'environment'. The imperatives of changing times reflected in the socioeconomic and ecological phenomena tend to make it rational on the national and international communities to address these issues. A static norm cannot hold for a very long horizon in a dynamic scenario, and if such insistence is sought, it tends to seriously expensive – thus defeating the purpose of collective action for common good.

The interpretation of like products in Article III.2 of the GATT limits the scope of national governments to differentiate products for environmental protection purposes.

Let us consider a few examples of the application of national laws in governing trade, especially import trade. The EC in 1991 imposed a ban, to take effect in 1996, to prohibit imports of certain furs from any country where the 'leg-hold trap' is used. This was considered relevant to bring about 'humane trapping standards'. In the US, the 1971 Pelly Amendment to the Fishermen's Protection Act 1967 authorized the President to prohibit the importation of any product from a country which allowed fishing operations that 'diminish the effectiveness of an international fishery conservation program' or engage in trade that 'diminishes the effectiveness of any international program for endangered or threatened species'. In 1993, Norway was certified by the US Commerce Secretary as its resumption of commercial hunting of mink whales attracts the application of the Pelly Amendment. Similarly, in 1994 the US President proposed to prohibit the importation of all wildlife products from Taiwan because the sale of rhinoceros horns and tiger bones in Taiwan was undermining the CITES. It is useful to note that CITES itself allows that it 'shall in no way affect the right of Parties to adopt...stricter domestic measures regarding the conditions for trade'.

The 1972 Marine Mammal Protection Act (MMPA), and its later modifications of 1988 led to a ban on the imports of tuna from some of the regions including Mexico. In connection with the application of a domestic US law to the international trade aspects, two GATT Panels held the US ban on imports of tuna from specified countries violated GATT Article XI, and that Article XX cannot come to provide relevant exceptions to the rest of the GATT Articles in this case. The EC challenged the application of the MMPA for trade and another GATT Panel of 1994 ruled against the GATT.

It appears that the GATT Panels used rather feeble arguments and dubious reasoning (see also Chang, 1995) in the interpretation of Article XX of the GATT. The 1991 Panel Report asserted that (a) '(a) country can effectively control the production or consumption of an exhaustible natural resource only to the extent that the production or consumption is under its jurisdiction'; and (b) 'Article XX (g) was intended to permit contracting parties to take trade measures primarily aimed at reducing effective restrictions on production or consumption within their jurisdiction'. As Chang (1995) argued, the Panel held implicitly that the natural resource itself must be within the territorial jurisdiction of the country employing trade measures, and the Panel's deployment of 'effective' measure is not supported by the text of the Article XX (g) which merely required the trade measure to be 'primarily aimed at rendering domestic restrictions effective'.

The 1994 GATT Panel on Tuna–Dolphin held that the MMPA violated GATT Article XI. This Panel rejected the 'extra jurisdictional' rationale of the 1991 Panel decision, but ruled against the US, using the following argument (GATT, 1994, para 5.27): 'measures taken so as to force other countries to change their policies, and that were effective only if such changes occurred, could not be primarily aimed either at the conservation of an exhaustible natural resource, or at rendering effective restrictions on domestic production or consumption, in the meaning of Article XX (g)'.

This Panel did also argue the following (GATT, 1994, para 5.26):

> If … Article XX were interpreted to permit contracting parties to take trade measures so as to force other contracting parties to change their policies …, the balance of rights and obligations among contracting parties, in particular the right of access to markets, would be seriously impaired. Under such an interpretation of the General Agreement could no longer serve as a multilateral framework for trade among contracting parties.

Whereas the fundamental requirement of non-discrimination should remain in tact, it is futile to seek its extended application to the detriment of the environmental and hence economic sustainability itself. The GATT 1994 under the new WTO framework should provide pragmatic measures which recognize the complementarity of trade and environmental measures. Rather than playing the role of a reluctant partner in the global integration issues, the WTO/GATT mechanisms should play a more responsible role with the adoption of new reforms and changes in some of the provisions under the respective articles of agreement. Some of the required specific measures in this regard are presented in Chapters 7 and 8.

When are the ETMs likely to be suitable for attaining international environmental objectives? Given below are set of prerequisites for utilizing ETMs in the governance of global trade and environment. These are not sufficient to ensure the pattern of sustainable development, however. Some of the elements below parallel those suggested in the report USOTA (1992). A useful debate on the ETMs is also provided by Charnovitz (1993).

ETMs are specially important instruments in any of the following scenarios:

1 those involving international externalities like ozone depletion or accumulation of greenhouse gases;
2 lack of workable incentives for administrative or market-oriented interventions;
3 transaction costs of achieving these environmental measures with the exclusion of ETMs exceed those with their inclusion (in conjunction with other relevant complementary programmes);
4 countries participating in the measures (or WTO members) do not themselves practise any less environment-friendly production and consumption methods than those sought under multilateral ETMs; and
5 there exists a strong positive correlation with ETMs and environmental benefits, and the benefits of these measures outweigh the costs of deploying the same, with particular reference to global economic welfare losses as result of the proposed ETMs.

5.5 Concluding observations

Policy developments from the WTO for the integration of environmental and trade factors are essential; those could make an earthly difference between environmentally sustainable trade and unsustainable free trade. The progress in this direction has been tardy during the 5 years that followed the formation of the WTO. Most of the existing provisions of the Articles of the WTO/GATT framework are obsolete as far as the international law, science, and economics of the environment are concerned. There is cause for concern in so far as the WTO Articles are oblivious to the implications of specific types of trade activities on the global commons and the implied free rider problems. Even if some trade activities are not the main contributors of the environmental problems, the need for an equitable assignment of the costs of production and trade inclusive of appropriate environmental costs is necessary between various members of the WTO. In the absence of such a level playing field and of a concerted action in this regard, individual countries may either attempt to free ride or be unwilling to include such costs for fear of loss of the market share. Alternately, the WTO Articles would bring about an actionable intervention on such members. None of these is conducive for harmonizing trade and the environment. Greater participation of the World Bank and the IMF to integrate the issues of including debt servicing and environmental issues, is essential. This integration could be sought in relation to debt and trade as well as trade and environment linkages. This is particularly relevant if these institutions can appreciate the role of integrating sovereign debt contracts with the compliance mechanisms required of various relevant multilateral environmental agreements.

Appendix I
Committee on trade and environment

(Decision of 14 April 1994)
GATT Ministers, meeting on the occasion of signing the Final Act embodying the results of the Uruguay Round of Multilateral Trade Negotiations at Marrakesh on 15 April 1994, proposed to the new WTO General Council to constitute a Committee on Trade and Environment to:

(a) to identify the relationship between trade measures and environmental measures, in order to promote sustainable development;

(b) to make appropriate recommendations on whether any modifications of the provisions of the multilateral trading system are required, compatible with the open, equitable and non-discriminatory nature of the system, as regards, in particular: the need for rules to enhance positive interaction between trade and environmental measures, for the promotion of sustainable development, with special consideration to the needs of developing countries, in particular those of the least developed among them; and the avoidance of protectionist trade measures, and the adherence to effective multilateral disciplines to ensure responsiveness of the multilateral trading system to environmental objectives set forth in Agenda 21 and the Rio Declaration, in particular Principle 12; and surveillance of trade measures used for environmental purposes, of trade-related aspects of environmental measures which have significant trade effects, and of effective implementation of the multilateral disciplines governing those measures;'

The terms of reference of the Committee on Trade and Environment were described in the Decision as follows:

that, within these terms of reference, and with the aim of making international trade and environmental policies mutually supportive, the Committee will initially address the following matters, in relation to which any relevant issue may be raised;

the relationship between the provisions of the multilateral trading system and trade measures for environmental purposes, including those pursuant to multilateral environmental agreements;

the relationship between environmental policies relevant to trade and environmental measures with significant trade effects and the provisions of the multilateral trading system;

the relationship between the provisions of the multilateral trading system and: (a) charges and taxes for environmental purposes; (b) requirements for environmental purposes relating to products, including standards and technical regulations, packaging, labeling and recycling;

the provisions of the multilateral trading system with respect to the transparency of trade measures used for environmental purposes and environmental measures and requirements which have significant trade effects;

the relationship between the dispute settlement mechanisms in the multilateral trading system and those found in multilateral environmental agreements;

the effect of environmental measures on market access, especially in relation to developing countries, in particular to the least developed among them, and environmental benefits of removing trade restrictions and distortions;

the issue of exports of domestically prohibited goods;

and that the Committee on Trade and Environment will consider the work program envisaged in the Decision on Trade in Services and the Environment and the relevant provisions of the Agreement on Trade-Related Aspects of Intellectual Property Rights as an integral part of its work, within the above terms of reference.

Appendix II
Extract of the SPS Agreement

Desiring to improve the human health, animal health and phytosanitary situation in all members, ... , to elaborate rules ... of GATT ... in particular the provisions of Article XX (b) provides the following:

Article 1: General provisions

1.1 This Agreement applies to all sanitary and phytosanitary measures which may, directly or indirectly, affect international trade. Such measures shall be developed and applied in accordance with the provisions of this Agreement.

1.2 For the purposes of this Agreement, the definitions provided in Annex A shall apply.

1.3 The annexes are an integral part of this Agreement.

1.4 Nothing in this Agreement shall affect the rights of Members under the Agreement on Technical Barriers to Trade with respect to measures not within the scope of this Agreement.

Article 2: Basic rights and obligations

2.1 Members have the right to take sanitary and phytosanitary measures necessary for the protection of human, animal or plant life or health, provided that such measures are not inconsistent with the provisions of this Agreement.

2.2 Members shall ensure that any sanitary or phytosanitary measure is applied only to the extent necessary to protect human, animal or plant life or health, is based on scientific principles and is not maintained without sufficient scientific evidence, except as provided in paragraph 7 of Article 5.

2.3 Members shall ensure that their sanitary and phytosanitary measures do not arbitrarily or unjustifiably discriminate between Members where identical or similar conditions prevail, including between their own territory and that of other Members. Sanitary and Phytosanitary measures shall not be applied in a manner which would constitute a disguised restriction on international trade.

Article 5: Assessment of risk and determination of the appropriate level of sanitary or phytosanitary protection

5.1 Members shall ensure that their sanitary or phytosanitary measures are based on an assessment, as appropriate to the circumstances, of the risks to human, animal or plant life or health, taking into account risk assessment techniques developed by the relevant international organizations.

5.2 In the assessment of risks, Members shall take into account available scientific evidence; relevant processes and production methods; relevant inspection, sampling and testing methods; prevalence of specific diseases or pests; existence of pest- or disease-free areas; relevant ecological and environmental conditions; and quarantine or other treatment.

5.3 In assessing the risk to animal or plant life or health and determining the measure to be applied for achieving the appropriate level of sanitary or phytosanitary protection from such risk, Members shall take into account as relevant economic factors: the potential damage in terms of loss of production or sales in the event of the entry, establishment or spread of a pest or disease; the costs of control or eradication in the territory of the importing Member; and the relative cost-effectiveness of alternative approaches to limiting risks.

5.4 Members should, when determining the appropriate level of sanitary or phytosanitary protection, take into account the objective of minimizing negative trade effects.

5.5 With the objective of achieving consistency in the application of the concept of appropriate level of sanitary or phytosanitary protection against risks to human life or health, or to animal and plant life or health, each Member shall avoid arbitrary or unjustifiable distinctions in the levels it considers to be appropriate in different situations, if such distinctions result in discrimination or a disguised restriction on international trade. Members shall cooperate in the Committee, in accordance with paragraphs 1, 2 and 3 of Article 12, to develop guidelines to further the practical implementation of this provision. In developing guidelines, the Committee shall take into account all relevant factors, including the exceptional character of human health risks to which people voluntarily expose themselves.

Article 7: Transparency

Members shall notify changes their sanitary or phytosanitary measures and shall provide information on their sanitary or phytosanitary measures in accordance with the Provisions of Annex B.

Annex A: Definitions

1. Sanitary and phytosanitary measures

Any measures applied:

(a) to protect animal or plant life or health within the territory of the Member from risks arising from the entry, establishment or spread of pests, diseases, disease-carrying organisms or disease-causing organisms;

(b) to protect human or animal life or health within the territory of the Member from risks arising from additives, contaminants, toxins or disease-causing organisms in foods, beverages or feedstuffs;

(c) to protect human life or health within the territory of the Member from risks arising from diseases carried by animals, plants or products thereof, or from the entry, establishment or spread of pests; or

(d) to prevent or limit other damage within the territory of the Member from the entry, establishment or spread of pests. Sanitary or phytosanitary measures include all relevant laws, decrees, regulations, requirements and procedures including, inter alia, end product criteria; processes and production methods; testing, inspection, certification and approval procedures; quarantine treatments including relevant requirements associated with the transport of animals or plants, or with the materials necessary for their survival during transport; provisions on relevant statistical methods, sampling procedures and methods of risk assessment; and packaging and labeling requirements directly related to food safety.

2. Harmonization

The establishment, recognition and application of common sanitary and phytosanitary measures by different Members.

References

Chang, H. F. (1995) 'An economic analysis of trade measures to protect the global environment', *Georgetown Law Journal*, 83.6, 2131–2213.
Charnovitz, S. (1993) 'A taxonomy of environmental trade measures', *Georgetown International Environmental Law Review*, 6, 1–25.
GATT (1994) *GATT Panel Report on Tuna–Dolphin*, Geneva: GATT Secretariat.

Ingersent, K. A., A. J. Rayner and R. C. Hine (1995) 'Ex-post evaluation of the Uruguay Round agriculture agreement', *The World Economy*, 18.5, 707–28.

Rao, P. K. (1999) *Sustainable Development: Economics and Policy*, Oxford and Boston: Blackwell Publishers.

Schoenbaum, T. J. (1997) 'International trade and protection of the environment the continuing search for reconciliation', *American Journal of International Law*, 91.2, 268–313.

US Office of Technology Assessment (USOTA) (1992) *Trade and the Environment – Conflicts and Opportunities*, Report #OTA-BP-ITE-94, Washington, DC: US Government Printing Office.

WTO (1999) *Background Note – Brief History of the Trade and Environment Debate in GATT/WTO*, prepared by the WTO Secretariat for the March 1999 High Level Symposium on Trade and Environment, Geneva: WTO Secretariat.

6
Dispute Resolution Mechanisms

6.1 Introduction

Perhaps the most significant improvement of the WTO over the GATT 1947 is its dispute settlement understanding (DSU). The GATT clauses for trade disputes settlement were too vague and too brief: contained in three precious paragraphs. The DSU, spelled out in 40 pages, incorporates several new dimensions such as the role of rules that cannot be vetoed by a single member country, clear schedules and time-bound actions in processing disputes and their effective compliance, recognition of the role of customary international public law and provision of measures to compensate members for adverse impacts of any trade policy violation by one or more measures. The DSU was expected to contribute toward security and predictability to the multilateral trading systems.

During the first four years of existence of the new DSU, the complaints by developed country members added to 124 requests dealing with 96 different subjects. Complaints by developing country members during this period were 31 requests, dealing with 29 different subjects. There were 4 complaints by both developed and developing country members with 10 requests – these included three requests made by developed country members. The total number of consultation requests made by developing country member was 40; on the basis of distinct matters, the number was 32.

The role of the global trading institutions and specifically that of the WTO is critical as well in this context. The US has been the most active litigant, both in filing cases against other members and in defending various cases filed against it. Some of the cases are rather insignificant in the context of global trade volumes affected or potentially to be affected, but the dispute is one of the compliance with agreed upon principles of trade and various rules evolved under the WTO Agreement. For example, five cases were filed a few months ago by the US against Belgium, France, Greece, Ireland and the Netherlands; all these relate to income tax subsidies, apparently in possible violation of the agreements on subsidies inhibiting competitiveness of exports. In the case against Belgium, this involved the provision of tax rules which enabled corporate tax payers to receive special tax exemption for recruiting personnel with export-related work functions.

This chapter highlights some of the important features of the DSU, and practical cases dealing especially with an interface of trade and environmental agreements. Appendix I summarizes the dispute settlement time table laid out under the DSU. One of the detailed case studies deals with the 8-year long US–EU banana dispute, given in Appendix II. Although this case does not fall under trade-environment disputes, the practical working of the GATT/WTO systems is elucidated with this illustration. Lessons of experience in this regard are particularly useful if the environmental trade measures possess an element of urgency and/or irreversibility. A few suggestions for improving the DSU are proposed towards the end of the chapter.

6.2   Legal foundations for dispute resolution

GATT Articles I (MFN), III (national treatment of 'like products'), and XI (limitations on the use of non-tariff barriers) are among the most significant provisions which could conflict with some of the proposed ETMs initiated at country level. These measures are generally handled at the multilateral level rather than unilateral levels. None of the three articles mentioned above warrants any significant changes in order to accommodate environmental considerations.

The original GATT process for dispute resolution had very little merit in its approaches and much less in dealing with any environmental trade measure-related disputes. Lack of provisions regarding the environment, coupled with almost total non-existence of environmental expertise on the GATT panels precipitated a series of rulings which were very narrow in their interpretation; most did not stretch beyond the interpretation of the words of the GATT Articles, possibly in a contextual manner. As Jackson (1998b) pointed, the flaws of the GATT dispute resolution process originated in the 'birth defects', including the 'sparse language with little detail about goals or procedures', and 'fragmented settlement procedures'. The DSU under WTO established a unified dispute settlement system for all parts of the GATT/WTO system.

Annex 2 of the WTO Agreement contains the Rules and Procedures Governing the Settlement of Disputes; some of the important schedules of disputes settlement process and the transmission of information for resolution of disputes are summarized in Appendix I of this chapter.

Two key WTO Articles of Agreement state:

1 Article XVI.1: 'the WTO shall be guided by the decisions, procedures and customary practices followed by the Contracting Parties to GATT 1947'.
2 Article XVI.3: 'In the event of a conflict between a provision of this Agreement and a provision of any of the Multilateral Trade Agreements, the provision of this Agreement shall prevail to the extent of conflict.'

As a general rule, the WTO dispute settlement stated in Article 3.2 of the DSU that the disputes will be resolved in accordance with 'customary rules of interpretation of public international law'. The GATT provisions and its environmental exceptions Article XX remain largely in tact under the new GATT 1994; several features of the role, merits and limitations of these provisions discussed in Chapters 3 and 5 (in addition to others) are relevant here. Besides, the overarching WTO Articles of Agreement, and other important Agreements like SPS, SCM, TBT and their interdependent interpretations constitute the legal bases for resolution of trade/trade-environment disputes under the WTO system.

The Dispute Settlement Body (DSB) of the WTO is the apex level group to offer a ruling on all trade disputes of the members. The DSU of the WTO, agreed by all the member countries of the world body states in its Article 21.1: 'Prompt compliance with recommendations or rulings of the DSB is essential in order to ensure effective resolution of disputes to the benefit of all Members.' The DSU was officially declared by the members as a 'central element in providing security and predictability to the multilateral trading systems'. According to the DSU, there is an obligation to perform and comply on behalf of the lost defendant in any dispute.

Article 16 of the DSU prevents blockage by a single member the DSB's acceptance of a panel report on any trade dispute. A panel report on a dispute will be deemed adopted unless there exists a consensus against it. This is called 'reverse consensus', a feature that constitutes a major improvement over the corresponding feature under the GATT mechanisms for dispute settlement. Unless a disputant opts to appeal, the panel report is submitted to the DSB for its consideration; the report is automatically adopted unless rejected by a consensus of the DSB. Appeals are permitted and can last up to 90 days. As pointed out in Jackson (1998a, b), a separate set of procedures for handling 'violation' and 'nonviolation' cases is incorporated for the first time for a treaty text.

Jackson (1997) summarized the operative parts of the DSU: the DSU establishes a preference for an obligation to perform the recommendation; indicates that compensation shall be required if the immediate withdrawl of the offending measures is infeasible or a non-event; and provides that in nonviolation cases, there exists no obligation to withdraw a disputed action, with an implication that in violation cases there exists an obligation to perform.

There has been a widely held view that the WTO mechanisms for dispute settlement are more legalistic than most other similar international institutions. However, the founding members, especially Canada, EC, and the US were in favour of such a system – with the intention of devising effective transparent methods of dispute resolution and averting any global effects of uncertainties in the trading regimes. The DSU in its Article 3.7 states that a 'solution mutually acceptable to the parties to a dispute and consistent with the covered agreements is clearly to be preferred (over the use of panel procedures)'. However, some parties tend to ignore this principle of wisdom and seek to test the workability of legal and institutional arrangements (see Appendix II regarding an 8-year-long banana trade dispute).

The next section summarizes a set of trade-related environmental disputes processed under the GATT/WTO systems during the past few years. As in any case law, the original legal provisions are then illuminated in the application of the law. Lessons of experience are useful for appropriate legal reforms and and any revisions in the original articles of agreement.

6.3 Trade-related environmental disputes in GATT/WTO

GATT hardly gained, nor even claimed any effective integration of trade policies and environmental considerations. Even after the 1992 UN Conference on Environment and Development (Rio Conference), it was clarified (GATT, 1992):

> the GATT's competence was limited to trade policies and those trade-related aspects of environmental policies which might result in significant trade effects for GATT contracting parties. In respect neither of its vocation nor of its competence was the GATT equipped to become involved in the tasks of reviewing national environmental priorities, setting environmental standards or developing global policies on the environment.

The narrow focus of the GATT mandate effectively precluded the incorporation of MEAs in trade polices. It is ironical that despite such a background under the GATT regime, a recent statement of the WTO Director General (WTO, 1999a) seeks environmental measures be the concern of the MEAs and would rather let WTO keep defending liberalized trade policies for their own sake.

Under the GATT, six panel proceedings involving an examination of environmental measures or human health-related measures under Article XX were completed. Of the six reports, three have not been adopted. Under the WTO DSU, two such proceedings have been completed. The following provides a factual overview of a select set of these disputes, based on the document WTO (1999b).

1 Canada – Measures Affecting Exports of Unprocessed Herring and Salmon, adopted on 22 March 1988

Canada maintained that under its 1976 Canadian Fisheries Act, regulations prohibiting the exportation or sale for export of certain unprocessed herring and salmon. The US complained that these measures were inconsistent with GATT Article XI. Canada maintained that these export restrictions were part of a system of fishery resource management aimed at preserving fishery stocks, and hence justified under Article XX(g).

The GATT Panel found that the measures maintained by Canada were contrary to GATT Article XI:1 and were justified neither by Article XI:2(b) nor by Article XX(g).

2 Thailand – Restrictions on Importation of and Internal Taxes on Cigarettes, adopted on 7 November 1990

Under its 1966 Tobacco Act, Thailand prohibited the importation of cigarettes and other tobacco preparations, but authorized the sale of domestic cigarettes. Besides, cigarettes were subject to an excise tax, a business tax and a municipal tax. The United States complained that the import restrictions were inconsistent with GATT Article XI:1, and considered that they were not justified by Article XI:2(c), nor by Article XX(b). The United States also requested the Panel to find that the internal taxes were inconsistent with GATT Article III:2. Thailand argued that the import restrictions were justified under Article XX(b) because the government had adopted measures which could only be effective if cigarettes imports were prohibited and because chemicals and other additives contained in US cigarettes might make them more harmful than Thai cigarettes.

The Panel found that the import restrictions were inconsistent with Article XI:1 and not justified under Article XI:2(c). It further concluded that the import restrictions were not 'necessary' within the meaning of Article XX(b). The internal taxes were, however, found to be consistent with GATT Article III:2.

3 United States – Taxes on Automobiles, not adopted, circulated on 11 October 1994

Three US measures on automobiles were under scrutiny: the luxury tax on automobiles ('luxury tax'), the gas guzzler tax on automobiles ('gas guzzler'), and the Corporate Average Fuel Economy (CAFE) regulation. The European Community complained that these measures were inconsistent with GATT Article III and could not be justified under Article XX(g) or (d).

The Panel found that both the luxury tax – which applied to cars sold for over $30,000 – and the gas guzzler tax – which applied to the sale of automobiles attaining less than 22.5 miles per gallon (m.p.g) – were consistent with Article III:2 of GATT.

The CAFE regulation required the average fuel economy for passenger cars manufactured in the US or sold by any importer not to fall below 27.5 m.p.g. Companies that were both importers and domestic manufacturers had to assess average fuel economy separately for imported passenger automobiles and for those manufactured domestically. During the first month of operation of the WTO, Venezuela complained to the DSB that the US was using discriminatory trade rules against gasoline imports. A year later the dispute panel completed its report. The panel and the DSB ruled that the US policy violated 'national treatment' provision under GATT and the GATT Article XX does not apply. The Panel found the CAFE regulation to be inconsistent with GATT Article III:4 because the separate foreign fleet accounting system discriminated against foreign cars and the fleet averaging differentiated between imported and domestic cars on the basis of factors relating to control or ownership of producers or importers, and not on the basis of factors directly related to the products as such. The Panel also found that the separate foreign fleet accounting was not justified under Article XX(g). The Panel found that the CAFE regulation could not be justified under Article XX(d). The US agreed and amended its regulations at the end of the implementation period in August 1997.

4   United States – Import Prohibition of Certain Shrimp and Shrimp Products, adopted on 6 November 1998

Seven species of sea turtles exist with their habitats distributed around the world in subtropical and tropical areas. Sea turtles have been adversely affected by their harvesting and also incidental capture in fisheries, destruction of their habitats, pollution of the oceans. In early 1997, India, Malaysia, Pakistan and Thailand brought a joint complaint against a ban imposed by the US on the importation of certain shrimp and shrimp products.

The US Endangered Species Act of 1973 (ESA) listed as endangered or threatened the five species of sea turtles that occur in US waters and prohibited their harvesting within the US, in its territorial sea and the high seas. Also, the US required that US shrimp trawlers use 'turtle excluder devices' (TEDs) in their nets when fishing in areas where there is a significant likelihood of encountering sea turtles. In practice, countries having any of the five species of sea turtles within their jurisdiction and harvesting shrimp with mechanical means had to impose on their fishermen requirements comparable to those borne by US fisherman in the shrimp harvesting activity, mainly to use of TEDs at all times, if they sought to be certified and need market access to export shrimp products to the US.

The Panel considered that the ban imposed by the US was inconsistent with GATT Article XI and could not be justified under GATT Article XX. The Appellate Body found that the measure qualified for provisional justification under Article XX(g), but failed to meet the requirements of the chapeau of Article XX, and was not justified under Article XX of GATT 1994.

One of the most significant and comprehensive analyses of environmental trade disputes relates to the tuna–dolphin issue, and this is explained in the following section in detail.

6.4   The Tuna–Dolphin dispute

The tuna–dolphin case still attracts considerable attention because of its implications for environmental disputes. It was handled under the old GATT dispute settlement procedure. Tuna often swim beneath schools of dolphins. When tuna is harvested with purse seine nets, dolphins are trapped in the nets. They often die unless they are released. The USMMPA sets dolphin protection standards for the domestic American fishing fleet and for countries whose fishing boats catch yellow fin tuna in that part of the Pacific Ocean. If a country exporting tuna to the United States cannot prove to US authorities that it meets the dolphin-safe standards set out in the US law, the US government must embargo imports of the fish from that country. In this dispute, Mexico was the exporting country concerned. Its exports of tuna to the US were banned. Mexico complained in 1991 under the GATT dispute settlement procedure. Tuna–Dolphin I: United States – Restrictions on the Imports of Tuna, not adopted, circulated on 3 September 1991.

The MMPA required a general prohibition of the 'taking' (harassment, hunting, capture, killing or attempt thereof) and importation into the US of marine mammals, except with explicit authorization. It governed, in particular, the taking of marine mammals incidental to harvesting yellow fin tuna in the Eastern Tropical Pacific Ocean (ETP), an area where dolphins are known to swim above schools of tuna. Under the MMPA, the import of commercial fish (or fish products) which have been caught with commercial fishing technology which results in the incidental kill or incidental serious injury of ocean mammals in excess of US standards were prohibited. Imports of tuna from countries purchasing tuna from a country subject to the primary nation embargo are also prohibited (intermediary nation embargo).

Mexico claimed that the import prohibition on yellow fin tuna and tuna products was inconsistent with Articles XI, XIII and III of GATT. The US maintained that the direct embargo was consistent with Article III and, in the alternative, was covered by Articles XX(b) and XX(g). The US also argued that the intermediary nation embargo was consistent with Article III and, in the alternative, was justified by Article XX (b), (d) and (g).

The Panel found that the import prohibition under the direct and the intermediary embargoes did not constitute internal regulations within the meaning of Article III, was inconsistent with Article XI:1 and also not justified by Article XX paragraphs (b) and (g). Also, the intermediary embargo was not justified under Article XX(d).

Mexico requested constitution of a panel in February 1991. A number of 'intermediary' countries also expressed their interest in the dispute. The panel reported to GATT parties in September 1991. It concluded:

- that the US could not embargo imports of tuna products from Mexico simply because Mexican regulations on the way tuna was produced did not satisfy US regulations. The panel's objection to process versus product distinction is clear;
- that GATT rules did not allow one country to take trade action for the purpose of attempting to enforce its own domestic laws in another country – even to protect animal health or exhaustible natural resources. The term used here is 'extra-territoriality'.

The US embargo was aimed at regulating not the sale of a product, but rather its process of production (the mode of harvesting in this case). Under GATT rules, the panel argued, the US was obliged to provide Mexican tuna (as a product) with a treatment no less favorable to that accorded to US tuna (also as a product), regardless of how the tuna itself was harvested. The panel also observed that whilst GATT Contracting Parties could adopt GATT-inconsistent measures, categorized under the 'General Exceptions' provision of GATT Article XX for the protection of the environment or the conservation of exhaustible natural resources, it was not clearly spelled out in the Agreement whether the resources being protected could fall outside the jurisdiction of the Party adopting the environmental controls.

The dispute did not end here. The new phase followed with a new complaint, described below.

Tuna–Dolphin II: United States – restrictions of the imports of Tuna, not adopted, circulated on 16 June 1994

The EEC and the Netherlands complained that both the primary and the intermediary nation embargoes, enforced pursuant to the MMPA did not fall under Article III, were inconsistent with Article XI:1 and were not covered by any of]the exceptions of Article XX. The US considered that the intermediary nation embargo was consistent with GATT since it was covered by Article XX, paragraphs (g), (b) and (d), and that the primary nation embargo did not nullify or impair any benefits accruing to the EEC or the Netherlands since it did not apply to these countries. The embargo also applied to 'intermediary' countries handling the tuna en route from Mexico to the United States. Often the tuna is processed and packed in one of these countries. In this dispute, the 'intermediary' countries facing the embargo were Costa Rica, Italy, Japan and Spain, and earlier France, the Netherlands Antilles, and the United Kingdom. Others, including Canada, Colombia, and the Republic of Korea, and members of the Association of the Southeast Asian Nations, were also named as intermediaries.

The Panel found that neither the primary nor the intermediary nation embargo were covered under Article III, that both were contrary to Article XI:1 and not covered by the exceptions in Article XX (b), (g) or (d) of the GATT. It was generally felt that if the US arguments were accepted, then any country could ban imports of a product from another country merely because the exporting country has different environmental, health and social policies from its own. It was opined that this could lead to a possible 'flood of protectionist abuses', and conflict with the main purpose of the multilateral trading system, that is, achieve predictability through trade rules.

The panel's conclusion that another international treaty the CITES would be relevant to the interpretation of a GATT provision only when the treaty was ratified by all GATT parties does not appear to be consistent with Article 31 (3) (c) of the Vienna Convention on the Law of Treaties, argued Palmeter and Mavroidis (1998). This Convention states that 'any relevant rules of international law applicable in the relations between the parties' shall be taken into account in a dispute resolution. Article 30 of the Vienna Convention is important in the application of MEAs. Article 30 (3) states that when the disputing countries are signatories of two agreements, the later in time prevails. Also, Article 30 (4) (b) specifies that, as between a state party to both treaties and a state party to only one of the treaties, the treaty to which both are parties becomes binding. As explained in Palmeter and Mavroidis (1998), this means that, in the event that a WTO member is a signatory to an MEA and another member is not, the MEA could not affect the rights and obligations under the WTO Articles of Agreement: 'a treaty binds only the contracting parties and neither rights nor duties arise under a treaty for a state not a party'.

The GATT 1991 Panel deployed rather feeble and 'dubious reasoning' in its interpretation of GATT Article XX(g), as explained by Chang (1995):

It pointed to a previous GATT (1988) Panel's position that a trade measure must be 'primarily aimed at rendering effective' the 'restrictions on domestic production or consumption' in order to be 'made effective in conjunction with' those restrictions (GATT 1991 Panel Decision, supra note 2, para 5.31, quoting GATT, 1988). The Panel asserted that a 'country can effectively control the production or consumption of an exhaustible natural resource only to the extent that the production or consumption is under its jurisdiction' and concluded that 'Article XX(g) was intended to permit contracting parties to take trade measures primarily aimed at rendering effective restrictions on production or consumption within their jurisdiction.' It is important to note that the GATT (1988) Panel decision ensured the compatibility of restrictions with the purpose of the trade measure and not its effectiveness. The GATT 1991 Panel decision with reference to interpretation of 'effective' measures is neither supported by the text of the Article XX(g) nor by the position of the previous panel it cites.

Also, as stated by Esty (1994), the major lacuna with Article XX is that 'it makes the legitimacy of environmental regulation turn on what is produced, not how it is produced'. The irrational rigidity in seeking compliance with neutrality toward PPM continues to plague the GATT provisions in so far as these cannot contribute toward harmonization of trade and environmental policies.

6.5   Toward improved methods

GATT Article III mandates national treatment of 'like products': it requires that imports from the territory of any contracting party 'be accorded treatment no less favorable than that accorded to like products of national origin in respect to all laws, regulations and requirements affecting their internal sale'. This requirement that all like products be treated equally limits the flexibility of a country to differentiate on the basis of environmental impact of products (during their life-cycle or different stages of production and usage). If the provision of the Article allows for comparison of 'like products' in terms of their environmental effects, we enter into the arena of differentiation on the basis of process and production methods as well. It is widely feared that such a distinction could allow some of the countries to utilize the provision to bring about disguised non-tariff barriers in the name of the environmental features underlying the products. It is likely that some cases could arise, but this problem is minimized when the basis of restriction is sought to be scientific and transparent guidelines are issued by the WTO for this purpose. A meaningful provision of process and product methods, with a slight modification of Article XX and/or Article XI should largely serve the purpose of integrating environmental considerations into trade issues. The inertia of some of the organizations or members should not be allowed to cost the economy–environment integration at the global and regional levels.

TRIPS regulates process and production methods (PPM): if a producer in a TRIPS member country resorts to patented production methods without obeying the appropriate international patent laws and exports such products to other member countries, an importing country could take 'appropriate measures' to prohibit such imports. As suggested in Fletcher (1996), a differentiation on the basis of production methods already found its origin under the TRIPS agreement, and an extension of similar reasoning to ecodumping would make sense. The latter could build on the provisions of anti-dumping and TRIPS to accommodate environmental aspects of trade.

To recognize the experience and perspectives of the Tuna–Dolphin Panel II of 1994, it is important to delineate a three step analysis (Fletcher, 1996) to determine the applicability of environmental exceptions provided under GATT Article XX(b). These involve the determination of the following: 1. whether the policies in question are really designed to 'protect human, animal, or plant life'; 2. whether these are 'necessary' for the stated objective; 3. whether the measures lead to an arbitrary or unjustifiable discrimination between countries where similar conditions prevail. The Panel stated that the eligibility of trade measures 'necessary' was allowed only when other less GATT-inconsistent measures were exhausted. Logically, a panel can usually come with an unexploited measure as potential candidate for reject-ing any application of 'necessary' measures. When environmental considerations are important, such a ranking of alternatives can ignore the merits of these considerations and weigh the merits of alternatives simply in terms of their GATT-consistency or selection of the least GATT-inconsistent measure. These hazards can be avoided when the trade measures are evaluated with reference to GATT provi-sion – which is still lacking – that allows a comprehensive economic and environmental impact assessment of proposed ETMs, while main-taining the main principles of non-discrimination. Under the new proposal, an evaluation of an ETM is based on a weighted index of agreed upon objectives, follows an agreed pattern of PPM differentia-tion and other guidelines of adoption.

GATT Panel decisions are reflective of the position in the pream-ble to the GATT Articles, which states that GATT is meant to facili-tate the 'full use of the natural resources of the world'. Let us note that this prescription is vastly different from a required 'optimum' use of natural resources (leaving aside, for a moment, alternate defi-nitions and premises for optimum use). A full use scenario is more like that of 'there is no tomorrow', and is built on the disparate high discounting of future. A number of limitations in the useful applica-tions of the GATT Articles arose from such background. GATT Panels have clearly refused to recognize the application of provisions of various MEAs (see, for example, Tuna–Dolphin 1994 Panel Report supra note 70 at Section 5.19).

Under the GATT/WTO, DSB does not consider obligations of members under MEAs even when both disputants ratified the competing or complementing treaty, as was amplified in the Tuna–Dolphin 1994 Panel Report. The Panel refused to consider any obligations under CITES or other MEAs. The Panel cited Article 31 of the Vienna Convention with a possible misinterpretation of the requirement that all parties must have been signatories of both the agreements (GATT as well as an MEA like CITES). The reference here to 'all parties' should have been all disputants and not all signatories of all agreements under consideration for dispute resolution (for detailed explanations in this regard, see Fletcher, 1996).

Finally, it is important that the new GATT 1994 and the DSU recognize the prescriptions of the preamble to the WTO charter wherein the roles of the need to protect the environment, and the need for sustainable development are clearly stated. It remains an empty foundation if these aspects are not fully integrated into the DSU and its methods of dispute resolution.

Appendix I
The dispute settlement panel process

The various stages a dispute can go through in the WTO are given below (WTO, 1998). At all stages, countries in dispute are encouraged to consult each other in order to settle 'out of court'. At all stages, the WTO director-general can offer his good offices, to mediate or to help achieve a conciliation.

Consultations
60 days (Article 4)

Panel established
by DSB (Article 6)

Terms of reference (Article 7)
about 20 to 30 days Composition (Article 8)

Panel examination
Meetings with parties (Article 12)
and third parties (Article 10)

Interim review stage
Descriptive part of report sent to parties for comment (Article 15.1)
Interim report sent to parties for comment (Article 15.2)

Panel report issued to parties
3 to 6 months from panel formation (Article 12.8; Appendix 3 para 12(j))

Panel report circulated to DSB
within the next 3 months (Article 12.9; Appendix 3 par 12 (k))

DSB adopts panel/appellate report(s)
including any changes to panel report made
by appellate report, 60 days for panel report (Article 16.1, 16.4 and 17.14)

Implementation
report by losing party of proposed implementation
within 'reasonable period of time', up to 15 months (Article 21.3)

In cases of non-implementation
parties negotiate compensation
pending full implementation (Article 22.2)

Retaliation
If no agreement on compensation, DSB authorizes retaliation
pending full implementation, 30 days after 'reasonable period' expires
(Article 22)

Cross-retaliation:
same sector, other sector, other agreement (Article 22.3)

Appendix II
Trade disputes on bananas

The United States and also four of the Latin American countries Ecuador, Guatemala, Honduras and Mexico lodged their complaint in 1996 with the World Trade Organization against the trading practices of the EU. This dispute arose as the EU is believed to have been in the practice of providing preferential treatment to the import of bananas from some of the former colonies of the British and the French in Africa, the Caribbean, and the Pacific. The recent trade dispute involving the US and EU is founded on a set of rather complex issues. On the surface, however, it appears deceptively simple: it is centered around the hurdles to free trade and the market access for bananas from the United States or its companies operating in many of the Latin American countries. Some suggested (from either side of the Atlantic) that the US hardly exports much of the bananas grown from its soil, and yet a fully-pledged trade dispute at the global level and a potential trade war was initiated. This is a consequence of an alleged failure of EU to comply with the rulings of the WTO. The present trading regime and the banana episode suggests the precipitation of the resultant outcome of the following factors: excessive reliance on legal procedures in lieu of economic diplomacy, interface of the interests of trading blocs like the European Union

confronting non-bloc countries, and challenges to the potential legal loopholes of the new international arrangements under the WTO.

European Communities – Regime for the importation, sale and distribution of bananas, complaints by Ecuador, Guatemala, Honduras, Mexico and the United States (WT/DS27)

The period for implementation was set by arbitration at 15 months and
1 week from the date of the adoption of the panel reports, that is, 1 January
1999. The EC has revised the contested measures. On 18 August 1998, the
complainants requested consultations with the EC (without prejudice to
their rights under Article 21.5), for the resolution of the disagreement between
them over the WTO-consistency of measures introduced by the EC in
purported compliance with the recommendations and rulings of the Panel
and Appellate Body. At the DSB meeting on 25 November 1998, the EC
announced that it had adopted the second Regulation to implement the recommendations of the DSB, and that the new system will be fully operational
from 1 January 1999. On 15 December 1998, the EC requested the establishment of a panel under Article 21.5 to determine that the implementing
measures of the EC must be presumed to conform to WTO rules unless challenged in accordance with DSU procedures. On 18 December 1998, Ecuador
requested the re-establishment of the original panel to examine whether the
EC measures to implement the recommendations of the DSB are WTO-
consistent. At its meeting on 12 January 1999, the DSB agreed to reconvene
the original panel, and pursuant to Article 21.5 of the DSU, to examine both
Ecuador's and the EC's requests. Jamaica, Nicaragua, Colombia, Costa Rica,
Cot d' Ivoire, Dominican Republic, Dominica, St. Lucia, Mauritius and
St. Vincent indicated their interest to join as third parties in both requests,
while Ecuador and India indicated their third-party interest only in the EC
request. On 14 January 1999, the United States, pursuant to Article 22.2 of
the DSU, requested authorization from the DSB for suspension of concessions
to the EC. At the DSB meeting on 29 January 1999, the EC, pursuant to
Article 22.6 of the DSU, requested arbitration on the level of suspension
of concessions requested by the United States. The DSB referred the issue of
the level of suspension to the original panel for arbitration within 30 days.
Pursuant to Article 22.6 of the DSU, the request for the suspension of concessions by the United States was deferred by the DSB until the determination, through the arbitration, of the appropriate level for the suspension of
concessions.

Further developments

Having itself complied with a number of rulings under the multilateral trade
regimes governed by the WTO, it was not easy for the US to condone the defi-
ance of the EU to the DSU mechanism and its operative rules under the WTO.

For the record, some of the proclaimed intentions of the EU may be
recalled. The WTO undertakes Trade Policy Reviews for every member, once

in about 4 years. At the 1996 Review of the EU, the Representative of the EU noted: '…trade creation not trade diversion, is the objective and the outcome; and that preferential trade is permitted only where this contributes to the collective (multilateral) good'. The practice of some of the trading policies involving EU and a few other trading blocs is a different story, however. Regional trading blocs and institutions like EU tend to undermine the global needs of multilateralism in various trade regimes, if the global arrangements under the WTO do not address the potential loopholes appropriately. However, WTO members have undertaken several initiatives in trade policy and its operationalization on a regional basis more expeditiously than at the multilateral level. The relative ease with which these undertakings can be effected, and thus minimize potential transaction costs involved in the full spectrum of trading operations (including various uncertainties of policy and realization of commercial revenues at predictable rates and volumes) tend to promote regional arrangements. Though reasonably meaningful, their interface with the multilateral institutional arrangements pose severe problems, as depicted in the current banana trade regime.

The banana dispute also brings to light a number of related aspects of multilateral trade. These include the need for improved methods of consultative resolution of trade dispute rather than intricate legal exercises, clarifications about the interface between trading blocs and the WTO rules (especially those of the Rules of Origin about traded goods and services) and the role of assessment of transaction costs in the trade regimes. Solutions for settling disputes via bargaining mechanisms have not been attempted and legalistic procedural challenges were attempted by the defendant EU. Thus, the banana dispute does highlight the lacuna of some of the current institutional mechanisms governing the multilateral trading regimes.

The phenomenon of excessive reliance on legalism in trade dispute resolution was foreseen even in the early stages of the formation of the WTO Agreement. Under normal rationality expectations, a member country's perception of the image in the international economic diplomacy could act as a factor in the dispute resolution process. For the EU, this did not seem to count, despite the potential global perception that the trading bloc disobeyed the WTO and constituted the first major challenge to the multilateral trading regimes.

It seems that the EU took advantage of that weakness under the previous mechanism to block the adoption of two panel reports (in 1993 and 1994) on banana trade which ruled in favor of the US complaint. The new provisions under the DSU did seem to make a major difference in the settlement of trade disputes, until the current issue of bananas and the stand taken by the EU. Despite some of the objections by the EU that the US has no material interest or a proper standing in the case of banana regimes, the 1997 Appellate Body Report under the WTO observed: 'The United States is producer of bananas, and a potential export interest by the United States cannot be excluded. The internal market of the United States for bananas could be affected by the EC banana regime, in particular, by the effects of that regime on world supplies and world prices of bananas.'

A few milestones in the evolution of the now notorious banana dispute may be summed up. In May 1993 the GATT panel found the EC in violation of the provisions of the GATT, and in July 1993 EC blocs' adoption of the panel repeat; this cycle repeats itself again during January and February 1994 in respect of (rather in lack of respect of !) the second panel under the GATT. During 1995–96, soon after the formation of the new WTO, the US and four of the Latin American Countries file the dispute under the new DSU. In May 1997 a WTO panel finds violations in the EC banana trade regime, soon after which the EC appeals. In September 1997, the Appellate Body under the WTO upholds the panel findings and later, the DSB adopts the panel report. EC virtually ignores the recommendations and scope for negotiated settlement. In January 1998, WTO Arbitrator gives EC until 1 January, 1999 to comply with the WTO panel rulings. In June 1998, EC's Agricultural Council adopts a few minor modifications to the banana trade rules and unilaterally declares them to be WTO-consistent.

The retaliation option was available to the US when authorized by the DSB. The EU objected to the magnitudes of imposition of tariffs (rates and coverage of items), the matter can be proposed for reference to a mandatory arbitration to decide whether the level of retaliation is equivalent to the level of the offense ('nullification or impairment at stake', according to Article 22.7 of Annex 2 of the DSU of the WTO). This mechanism tends to limit the type and degree of retaliation. It goes without much confusion that any retaliation can be a lose–lose scenario and retrograde step in the ultimate analysis. The real lacuna in the current list of provisions under the WTO and its DSU seem to be lack of realistic compensatory measures to offset any trade effects of tainted trade measures adopted by members. There is no plausible reason why these measures cannot be devised and brought under the DSU. This can be accomplished at the next Ministerial Conference of the WTO.

US requested the WTO its authorization to impose trade tariffs and suspend concessions to specific items of import from EU (except those of the Netherlands and Denmark). The specific amounts was determined in April 1999 at $191 million, determined by a binding arbitration. It is apt to recall an observation of the 1997 Appellate Body Report of WTO on the EC banana import regime: '...with the increased interdependence of the global economy...Members have greater stake in enforcing WTO rules than in the past.'

References

Chang, H. F. (1995) 'An economic analysis of trade measures to protect the global environment', *Georgetown Law Journal*, 83.6, 2131–213.

Esty, D. C. (1994) *Greening the GATT – Trade, Environment, and the Future*, Washington, DC: Institute for International Economics.

Fletcher, C. R. (1996) 'Greening world trade – reconciling GATT and multilateral environmental agreements within the existing world trade regime', *Journal of Transnational Law and Policy*, 5.2, 341–72.

GATT (1988) *Report of the GATT Panel, Canada – Measures Affecting Exports of Unprocessed Herring and Salmon*, GATT Doc. L/6268 (para 4.6), Adopted 22 March, 1988.

GATT (1992) *GATT's Follow-Up to the UN Conference on Environment and Development*, GATT Doc. SR.48/1 (2 December 1992) (decision by Contracting Parties).

Jackson, J. H. (1997) 'The WTO dispute settlement understanding – misunderstandings on the nature of legal obligation', *American Journal of International Law*, 91.1, 60–4.

Jackson, J. H. (1998a) *The World Trade Organization – Constitution and Jurisprudence*, London: The Royal Institute of International Affairs.

Jackson, J. H. (1998b) 'Dispute settlement and the WTO – emerging problems', *Journal of International Economic Law*, 1.3, 329–51.

Palmeter, D. and P. Mavroidis (1998) 'The World Trade Organization legal system – sources of law', *American Journal of International Law*, 92.3, 398–413.

WTO (1998) *Trading Into the Future*, Geneva : WTO Secretariat.

WTO (1999a) *Director General's Opening Remarks at the WTO High Level Symposium on Trade and Environment*, Geneva: WTO Secretariat.

WTO (1999b) *Background Note – High Level Symposium on Trade and Environment*, Geneva: WTO Secretariat.

Part III
Policy Implications

7
International Trade and Environment: an Integration

7.1 Introduction

Some of the extreme positions like free trade for its own sake or an absolute environmental protection are unsustainable, the former because of its ignorance of linkages with economic and ecological systems and the latter because of its severe restrictions on economic growth, and both on sustainable development. In general, free trade rather than protected trade contributes to enhanced economic growth, but balanced trade and optimal trade (see Chapter 1 for definitions of various trade policy regimes) could bring about economic growth as well as attain other economic objectives. In the multilateral trading system, these could contradict some of the provisions of the WTO framework, however. Sustainable trade maintains a comprehensive approach toward sustained economic and environmental progress, recognizing the complementarity of both. Deriving operationally meaningful and WTO-consistent policy guidelines for an application of this conceptual framework remains a major task. This is the focus of this chapter.

The impact of trade and its liberalization may be positive, negative or null depending on the composition of the economic and environmental basket from which trade emerges in any given configuration, suggested an OECD (1994) report. It was argued: 'In general, trade is not the root cause of environmental problems, which are due to market and intervention failures.... International trade can help correct ... failures through increased funds and incentives for environmental protection and promote efficient resource use.

But, at times, … may exacerbate the environmental problems.' In general, the operationally significant issue whether freer trade regimes do indeed 'help correct market failures' and/or 'intervention failures', and whether these instead add their 'failures' to those already in existence.

After a brief review of general principles devised during recent years in multilateral agreements, a detailed examination of the relevance and applicability of major principles of environmental governance within the context of international trade is provided in the following sections. These include the relevance of Precautionary Principle and Polluter Pays Principle. The imperatives of recognizing and reflecting environmental costs in trade, and improvisation of mechanisms to address the adverse implications of 'market failures' constitute a major aspect of these analyses. Avoidance of costly conflicts between trade and environmental objectives, and development of cost-effective policies and operational means to enhance the complementarity of the two sets of objectives is the critical element of an exercise in the integration of trade and environmental policies.

7.2   Cost externalization and markets

It is useful to recall a classic definition of external costs (Kapp, 1963: 10): all direct and indirect losses which third parties or general public are forced to bear as result of economic activities; such social costs include damage to human health, loss of assets and the premature exhaustion of natural resources. The phenomena of ecological dumping (details are elucidated in Massarrat, 1997) are based on the processes of cost externalization. Externalization of social costs in early capitalism was well known. It might have led to capital formation in some cases but that was achieved at significant human costs. Intergenerational externalization of costs can occur with the neglect of long-term interests of future generations. This is sometimes described as a failure of the market to recognize the needs of the future, and also seen as a consequence of 'missing markets'. In either case, externalization has several components: producer externalizing environmental costs to others; consumer doing the same with disposal activities that exacerbate waste assimilation capacities of the sinks of the planet; exporter transmitting pollution contents for further processors who incur these direct and indirect costs; current generation exploiting the natural capital assets to the detriment of the interests of the future generations; developed countries affecting global terms of trade with greater control of capital and technology, and globalizing environmental costs; the weak and helpless bearing the costs of externalization caused by the powerful. Cost externalization (CE) is thus an unethical proposition, viewed from any perspective. The Principle of Polluter Pays (explained later in this chapter) is just an example of measures to avoid CE or minimize in a specific context of the environmental pollution costs. In its weaker form, it still admits some element of CE phenomenon and could provide perverse incentives for emissions of pollution. Let us recall that Agenda 21 of the 1992 Rio Conference recommends a good deal of minimization of CE when it stated (para 2.14(c)): 'efficient and sustainable use of factors of production in the formation of commodity prices, including the reflection of environmental, social and resource costs'.

The issue of internalization of environmental costs is but a corollary of minimization of cost externalization. To relate to much of neoclassical economic terminology, CE is an envelope of the set of externalities generated from a wider set of perspectives than simply market externalities or price externalities. It is also useful to note that internalization of environmental costs is not uncorrelated to the process of internalization of environmental benefits, only involving a time shift factor and spatial interactions of environmental inter dependencies. Thus, due recognition of environmental costs and their internalization leads to reaping of environmental benefits and thus these are reflected later in the economic systems in terms of sustainable and efficient production and consumption activities.

Environmental benefits of an efficient and sustainable resource use tends to retain the uninterrupted supply of services of nature and ecosystems, thus control costs and ensure supply of resources at reasonable transaction costs. However, the absence of functioning markets governing services of nature could disguise the effortless internalization of environmental benefits. Sustainable development does imply some of these operative features. Thus, it is very critical to the trade and other economic activities to utilize various principles including an extended version of the Polluter Pays Principle on a multilateral basis. As the UNCTAD (1995) report observed, internalization of environmental costs and benefits occurs within the different country-specific domestic economic and environmental policies, their development priorities, environmental absorptive capacities and time preferences involved.

When environmental resources are under priced, the short-term effects of full cost internalization could include an increase in production costs, and when transmitted into the pricing mechanisms, the higher prices may not help withstand competitiveness (except in special cases). Internalization could also imply more efficient use of scarce resources, recycling, and adoption of more efficient methods of production. The adaptation mechanisms on the demand side as well as supply side might warrant incrementalism in the internalization of environmental costs. Such an approach could be Pareto-welfare improving (i.e. enhancement of welfare of participants without bringing down the welfare of any) for the producers and the consumers. It is not always feasible to incorporate all the environmental costs without major disruptions in the provision of affordable goods and services. Besides, there is reasonable possibility of transforming utilization of goods and services to higher productivity levels so that these productivity gains can offset the depreciation of environmental assets. The real issue is these promise the potential but do not necessarily deliver the results. This draws a wedge between gainers and losers of the gainful economic activities, and again an element of CE emerges. Hence the need for a multilateral consensus on the types and magnitudes of environmental cost internalization that remain pragmatic in the existing system, without losing sight of the imperatives of sustainable development. In general, if costs or benefits of trade are externalized, economic systems produce sub optimal or inefficient patterns of goods and services and the trade pattern fails to maximize the economic welfare of all trading partners (see also Tuchband, 1995).

The issue of product cost externalization is not necessarily the converse of the issue of internalization of environmental cost. This is because the question of full product costs could include other factors, in addition to the environment. In the context of multilateral trade regimes, unilateral principles of cost internalization are not likely to be competitive, and their contrasts with direct financial subsidies are usually WTO-inconsistent. The next section addresses the role and applicability of the alternative methods of internalization of environmental costs.

7.3   Internalization of environmental costs

Internalization of environmental costs does mean all or a few of the following: inclusion of true worth or shadow price of each of the inputs into the production system, post-production costs to consumption stage, and also the costs of disposal at the terminal stage. This inclusion does imply costs which are incurred at different levels of the life cycle. Evidently, in many cases, it is not meaningful to include all costs at a single stage. A multi-stage inclusion of environmental costs is one of the direct methods, but a variety of alternatives can also be contemplated. An alternative perspective suggests assessment of the opportunity costs of resources being used in the life cycle of the product or service. In an operationally relevant setting the objective of internalization is to enable various economic entities to adjust prices and markets so as to achieve socially optimal consumption and production patterns (see also Rao, 1999). Any degree of adoption of this approach does warrant concerted policy and action at different levels of aggregation of human activities, if any free-rider problems and their corresponding externalities have to be avoided. If a few firms in a few sectors and locations reflect the environmental costs in their prices, the competitors could capture the market share and hurt the profitability (and hence their survival) of the adopters of the principle.

In a competitive world, it is well known that cost-plus method of pricing might tend to exclude some of the producer markets. This has implications for the uniformity or harmonizing environmental costs by all the players (at least with respect to specific commodities/goods and their substitutes) in the world market: those who seek to free ride on the environment should not have an incentive to do so by their lower pricing of goods and services to clinch the export market. Hence the need for a global policy coordination. Instruments for internalization have both income effects and substitution effects at different levels of the market. These effects tend to work in opposite directions, thus the net effect on exporter income may be positive, negative, or none, as it depends on the specific characteristics of the product and the market under consideration (UNCTAD, 1995). Internalization acts first on the supply side, and interact with substitution mechanisms at consumption and production levels.

The 1992 Rio Declaration Principle 16 (Declaration of the Earth Summit) stated: 'national authorities should endeavor to promote the internalization of environmental costs and the use of economic instruments, taking into account the approach that the polluter should, in principle, bear the costs of pollution, with due regard to the public interest and without distorting international trade and investment'. Let us recall that the Rio Summit Declaration sought to 'reflect efficient and sustainable use of factors of production in the formation of commodity prices, including the reflection of environmental, social and resources costs' (Agenda 21, para 2.14(c)). The Summit itself fell short on the issuance of any specific guidelines or policies in this direction. Besides, part of assertion in its Principle 16 was inconsistent with itself. It is important to recognize that some element of distortion of the patterns of international trade is rather essential if the existing distortions are to be corrected. That is the price we may have to pay in order to correct a wrong. The failure to recognize this imperative is to seek *status quo ante* and still dream about effecting relevant changes. Ineffective changes will, of course, lead to non-distortive impact on international trade. Disruption of the trading system will be harmful, but the imposition of one calculated 'distortion' (if at all) to correct another existing distortion need not be harmful. The real issue is to find cost effective methods of effecting meaningful changes, where the concept of cost is the generalized one. It includes (subject to adjustments for various overlapping elements) the following: producer and consumer costs, social and private costs, compliance costs and transaction costs, and environmental costs for the present and future generations.

A rather limited perspective of internalization and a set of recommendations emerged from a UNEP–World Bank Workshop of 1995 (UNEP, 1996). It was stated that internalization of environmental costs is the incorporation of external costs and benefits in the decision making calculus of producers and consumers, with the objective of steering their decisions toward socially desirable outcomes. Subsidies, resource depletion costs, and environmental damages unaccounted in the consumer and producer levels were considered external costs. It was suggested that to the 'extent internalization involves local externalities incurred during the production process, policies which have competitiveness implications should be compatible with WTO rules'. However, the real problem is not the inconsistency of product costs that internalize local negative externalities with adverse implications on maintaining competitiveness in the global markets, but just the opposite: that of externalization of local and non-local externalities.

Is the process of internalization an internationally relevant and significant one or is it simply a domestic trade policy of a given country? UNCTAD (1995) observed that internalization of environmental costs and benefits need to be achieved within the country-specific domestic policies, environmental absorptive capacities and time preferences involved. This may not be entirely valid, especially when there exist unassimilated emissions of pollution which is more of a global externality: CO_2 remains good example of this pollutant. Again, this points to the need for an international coordination mechanism, market-based (like emissions trading) or quota-based or target-based reduction of global environmental pollution. One of the ingredients of internalization of environmental costs is their reflection in the costs of producer prices. This can initially create problems in competitiveness, especially in the export markets. When prices of goods transacted in the international market do not properly reflect the social costs of production, they are effectively receiving a subsidy equal to the uncompensated environmental resource use, partly despoiled or lost in the process (see also Porter and Brown, 1991). Let us note that international trade of goods produced under environmentally unsustainable conditions is unsustainable, just as consumption based on such conditions is also unsustainable.

In a report, UNCTAD (1997) suggested that full cost internalization will rarely be optimal and that internalization should only be carried out up to the level where the incremental benefits of avoided environmental damages justify the incremental costs of environmental provisions. This general economic argument could be a step in the right direction, especially when all the costs and benefits are properly taken into account. This approach is not significantly different from the generally accepted (but largely ignored in practice) principle 'polluter pays', except that internalization of environmental costs method could apply all over the life-cycle of the product: polluter producer, non-polluter consumer, and other intermediaries involved in the sale and distribution. There may be need for setting a floor price specific to some of the commodities in international trade. This might facilitate provision of revenues so generated to augment investments in environmental and ecological conservation. However, administrative enforcement problems could be significant serious in the application of this principle. A more pragmatic approach was suggested by Elliot (1994): rather than seeking to front-end load all the life-cycle environmental costs on the producer, it is desirable to unbundle part of these costs so that only environmental impacts which can be dealt with at the production stage of a product are internalized at the level of the producer; the costs of environmental impacts incurred at subsequent stages of product life-cycle can be apportioned and internalized at the level of those who benefit directly from the consumption of products under reference. A case study of internalization of costs, reduction in environmental pollution, and implications on competitiveness is given in Box 7.1 for palm oil in Malaysia.

Box 7.1 Palm oil cost internalization – Malaysia

The following is an example of internalizing environmental costs in the international market context. In Malaysia, oil palm cultivation was encouraged in the 1970s and 1980s with the intention of reducing reliance on rubber export. Although Malaysia supplies about 80 per cent of the palm oil that enters the world market, refined palm oil has to compete with 16 other products in the world market of fats and oil (among which soybean oil is the closest substitute). Palm oil output expanded rapidly and accounted for about 40 per cent of the increase in agricultural output during the 1980s. In the same period, however, the palm oil processing industry was responsible for more than 60 per cent of Malaysia's total water pollution load. The effluent caused serious depletion of dissolved oxygen and killed fish, prawns and crabs, which are important sources of nutrition and gainful employment.

Malaysia's refined palm oil sector lost only 5 per cent of the value of output, and the crude palm oil sector lost only about 1 per cent of the value of the world oils market. Internalization did not appear to have affected the palm oil industry. Processors could diversify most of the additional costs onto producers of

Box 7.1 continued

fresh fruit bunches (FFB), the planters and cultivators of oil palms. These producers appear to have borne 84 per cent of the total industry losses during the abatement period. Small holders and plantation owners were compensated to some extent by the provision of inexpensive fertilizers supplied as a by-product of effluent treatment. The palm oil industry was able to promote their by-products: fertilizers, animal feed, and energy generation via methane gases. The effluent standards contained in the environmental regulations were effective in the reduction of discharge. The role of the government in establishing in 1980 the Palm Oil Research Institute of Malaysia, and in funding relevant research is noteworthy. These findings demonstrate that internalization need not impair overall competitiveness; changes in the distribution of returns from trade do occur and these call for complementary measures to offset the externalities.

Sources: Rao (1999); Vainio (1998); UNCTAD (1994, 1995); Khalid and Braden (1993)

Although there is little policy specification under the WTO framework, a recent document (WTO, 1999) maintains the following: The 'basic premise' of the document is that:

> in well-functioning market-based economies, prices register the relative scarcity of resources and consumer preferences; their role is to, inter alia, allocate resources efficiently. The welfare of society can be undermined, however, when market prices fail to capture the effects of environmentally-damaging activities and therefore send misleading signals regarding the optimal use of environmental resources...Distorted prices obscure the abundance of under-utilized environmental resources, contribute to the excessive depletion of exhaustible resources, generate new environmental problems, and contribute to the excessive use of environmentally-damaging inputs.

Analytically, the problem of inclusion of environmental costs consists of maximizing the total of producer surplus and consumer surplus in various stages, with a provision for effecting income redistribution among the classes of economic agents involved in the total system. Such an approach could facilitate distribution of environmental costs in terms of efficiency and equity considerations, rather than simply in terms of stages of processing and their corresponding pollution costs. This is significantly different in its economic considerations compared to the Polluter Pays Principle (Rao, 1999): Environmental costs can be minimized with alternate forms of production, but direct costs of production may be higher than usual (possibly due to lower yields). One major difference, however, in the market place is that the conventional environment depleting (or pollution generating) methods of production allow the option of excluding environmental subsidies and thus pass on the externalities to the rest of the system. In most of the environment-enhancing methods of production, the direct unit cost of production tends to be higher and the producer has little option but to raise the selling price (assuming there are no fiscal incentives). In this case, the ecosystem services are subsidized to a lesser extent, and the costs are internalized to a greater degree than in the conventional production methods. Green preferences represent a new and rapidly growing dimension of trade and environment issues.

If a narrow view of environmental pollution is taken, for example, with the exclusion of greenhouse gases and CO, it may appear that developing countries have greater absorptive capacity in the financial and environmental regulation sense. The pollution-intensive industries do not necessarily specialize in these countries. Rather, the industries that exist are allowed to flourish even with severe environmental problems relative to those in the developed countries. But the externalities are transmitted via the features affecting the global commons, in addition to inflicting local pollution problems. This aspect may be illustrated below in the case of bilateral trade between Indonesia vs Japan.

The differences in the pollution intensity of trade between Indonesia and Japan was found in a study by Lee and Roland-Holst (1994). The intensity was measured by a human toxicity index of effluents, of production and trade. If a reference level for normalization (not because of the best exemplary system) is taken as one for the aggregative toxicity feature per unit of economic activity in the USA, Japan had an index of 0.86 and Indonesia an index of 2.45 showing the specialization in lesser clean technologies of production.

When it comes to trade, the pollution content of trade from Indonesia to Japan was 10.64 whereas that in the reverse direction was 1.62.

It is not hard to generalize a little on the basis of the above illustration: most of the developing countries are 'pollution havens', and various exports in the present form and practice are environmental subsidies of the poor to the rich: a blend of implicit and explicit subsidies of nature, women, and children of the poorer countries to the rich any where. Kox (1995) estimated that in developing countries, costs borne by the society as a result of environmental damage reaches up to about 17 per cent of GDP, and the corresponding estimates for developed countries are much lower. Internalization of environmental costs in exports, which may be more pollution intensive than other economic activities, could generally warrant a 15 to 20 per cent hike in prices. Selective full cost pricing is usually worse than gradual incremental and universal application of principles of internalization of environmental costs, possibly only partially. This is because of the recognition that the life-cycle approach warrants apportioning only part of the costs at the production front-end. International commodity-related environmental agreements (ICREAs) have been proposed as possible interim solutions to alleviate the international competitive pressures among the primary commodity producers, so as to enable exporting countries to pursue a gradual transition towards ecologically meaningful production methods, with an eventual phasing out (Kox, 1995).

As an illustration of the extent of costs which may need to be reflected in the prices of developing countries' exports, it was estimated that OECD members could have incurred direct pollution control costs of $5.5 billion for their 1980 imports from developing countries if they had been required to meet the environmental standards then prevailing in the United States (UNCTAD, 1995). If the pollution control expenditures associated with the materials that went into the final product had also been counted, the costs would have mounted to $14.2 billion in 1980 (Walter and Laudon, 1986). This should be considered in the light of the fact that industrial countries have generally been more successful than developing ones in passing into their export product prices the costs of environmental damage and of controlling that damage (UNCTAD, 1993). Thus, in the case of exports from industrial countries, consumers, including those in developing countries, bear at least part of the burden.

But in the case of exports from developing countries, the consequences of environmental damage are borne overwhelmingly by
domestic residents and firms, principally in the form of ill-health,
reduced productivity and higher costs. This asymmetry in product
pricing mechanisms is usually founded on the dire need for the
developing countries, who are also usually significantly indebted
countries, to earn foreign exchange. These motives of export and
the implied high discount rate on appropriation of environmental
resources (compounded by the myopic high discount rates on time
in case of some of the ruling regimes) remain a recipe for environmental disaster in some of the developing countries. In summary,
environmental costs which are not internalized, are effectively
internationalized with a multiplier effect (Rao, 1999). Detailed
environmental accounting and full cost pricing in each of these
cases is expected to lead to additional insight into the working
dynamics of the terms of trade, constrained exports for debt servicing, and environmental implications of such trade with implicit
subsidies.

It is possible that less transparent uses of trade restrictions can disrupt international trade relations, trade activities and global welfare,
insufficient to offset the merits of environmental protection potential or its realization. The capture of the usual tools of trade protectionist measure under the disguise of environmental protectionism
can be a net loss game for the importers and exporters. As Low
(1993) suggested, it is easy to get confused between using trade measures and other measures for environmental policies, without fully
realizing the relative ordering of priorities or their relative cost effectiveness in realizing the stated environmental objectives. Much
worse is the possibility of deploying the environment card to con-
fuse others in a deliberate game of availing such instruments to
devise protectionist policies of trade. Thus, it is important that
the environmental concerns of the expanding trade operations be
addressed but not by adopting unilateral or nontransparent methods. Such methods can only be self-defeating in economic and environmental objectives. The multilateral mechanisms must be allowed
to function effectively to the advantage of all countries, and in the
achievement of the goals of sustainable development.

Among the simpler mechanisms which partially reflect differential
environmental costs or inputs involves eco-labelling.

7.4 Ecolabels and their trade impacts

Ecolabels are double edged in their (positive or negative) contribution toward environmental enhancement with trade specifications. If carried out on a comprehensive scale, ecolabels amount to adoption of PPMs, a contentious issue in the existing GATT framework. Most objections raised by the Panels in the Tuna–Dolphin and several other environmental cases apply. The concern over PPMs goes on to the main element of eco-labelling programmes' reliance on life-cycle methodologies (Salzman, 1998) in as much as these refer to assimilative and recycling features of the products or the inputs used in those products. To the extent that the ecolabels may not involve the consent of all member countries of the WTO, they tend to be discriminatory and GATT-inconsistent (despite the suggestion to the contrary in the 1991 Tuna–Dolphin Panel report).

The following is an illustration based on Beghin *et al.* (1994) and Rao (1999). The Netherlands has a long tradition in horticultural exports such as cut flowers. Dutch dominance on world markets has been recently eroded, however, by the emergence of horticultural production and exports originating in developing economies. The flower industry is chemical-intensive and polluting both in the North and South. Dutch horticulture is also energy-intensive since most production occurs in greenhouses. A proliferation of green labels (for example, environmental standards defined by flower auction houses) has been occurring in the Dutch horticultural sector, partly to address growing environmental concerns and partly to differentiate their products from those produced by competitors in the South (for example, Colombia and Morocco). In 1995 the independent environmental foundation 'Milieukeur' awarded an independent eco-label for cut flowers considering five production stages and eight environmental dimensions that will mimic a cradle-to-grave approach. Importers can apply for the new label, but foreign producers will not be represented on the Milieukeur panels and are not involved in the definition of the label. It is unlikely that the two types of production (natural sunlight versus greenhouse) will be on equal footing. The emergence of the new standards ISO 14001 is likely to provide a common level playing field for most producers, if the standards are widely adopted.

Because of substitution effects, ecolabelling or other forms of credible certification can lead to decreased consumption and price of competition which by contrast is presumed environmentally un-friendly. The following is an example of growing area of differential income potentials for new environmentally conscious consumption markets which bring new avenues for producers, including those in the developing countries. Although the approach may not extend to every commodity or beyond a scale for applicable commodities, it does expand to augment trade and environmental complementari-ties to some extent with an element of sustainability. This presumes some stability or loyalty or consumer markets which have such tastes (which are often habit forming).

7.5   Integrating environment into subsidies agreement

The Agreement on SCM (called SCM Agreement) allows for 5 years from 1995 the provisions under Articles 6.1, 8 and 9 under the so-called dark amber and greenlight rules certain types of allowed sub-sidies. Let us recall that the definition of subsidy here is built on the simultaneous validity of a financial contribution by a government or any public entity within the territory of a member which confers a benefit. When a subsidy measure is not specific, in terms of the following, it is not subject to the Agreement and might not be 'actionable'. The four types of 'specificity' within the jurisdiction of the SCM Agreement are: enterprise-specific, industry-specific, region-specific, and specific to export goods/import substitution goods (or their related domestic inputs). The non-actionable subsi-dies fall into three groups, and these have implications for environ-mental considerations, explained below. (a) basic research and pre-competitive development subsidies (limited to a proportion of project costs and detailed criteria); (b) assistance to disadvantaged regions (using objective development criteria); and (c) assistance to adapt existing facilities to new environmental requirements, on a one-time basis and limited to 20 per cent of adaptation costs avail-able to all firms which can adopt the new equipment and processes. The Agreement also seeks to limit subsidies to 5 per cent of total production costs, in order to avoid 'serious prejudice' to the inter-ests of another member, and to allow only minimal trade distortion effects on production as a result of subsidies. The Agreement pro-vides a greater leverage for developing countries in their phase-out of actionable subsidies.

An international code for environmental policy somewhat similar to the SCM might be relevant. An important safeguard required is to provide guidelines distinguishing subsidies for environmental purposes from those with other motives (Siebert, 1996).

Agricultural subsidies are governed by a more general Agreement on Agriculture. In a submission (WT/CTE/W/106, 11 February 1999) to the WTO/CTE on 'Agriculture and the Environment – the Case of Export Subsidies' by Argentina, Australia, Brazil, Canada, Chile, Colombia, Indonesia, Malaysia, New Zealand, Paraguay, the Philippines, Thailand, United States, Uruguay maintained: Experience since the annual reduction commitments came into force in 1995 has shown wide variation in the extent to which WTO Members have notified export subsidies up to their annual limits (or beyond). Overall use of export subsidies has been less than half of the maximum level allowed under WTO rules, mainly due to relatively high world grain prices during this period and changes in agricultural policies in some Members.

7.6   The Precautionary Principle

The Precautionary Principle (PP) is based on the idea that any uncertainty should be interpreted toward a measure of safeguard. The PP is equivalent to 'No Regrets' policy. The Rio Declaration of the Earth Summit 1992 Agenda 21 Principle 15 stated: 'Where there are threats of serious or irreversible damage, lack of full scientific certainty shall not be used as a reason for postponing cost-effective measures to prevent environmental degradation ... In order to protect the environment, the precautionary approach shall be widely applied by States according to their capabilities.' Earlier, the Ministerial Declaration concluding the Second World Climate Conference in 1990 at Geneva stated exactly this principle. Some of the areas of concern and application of the PP should be in the area of loss of biodiversity resulting from of the trade in bioproducts, and the critical role of time element in resolving environment-related trade disputes.

The first international formulation of the Principle was at the First International Conference on the Protection of the North Sea in 1984 when the focus was on emissions into the marine environment. The PP played increasingly significant role since its endorsement by the Second International Conference of the North Sea in 1987.

The PP has been most frequently advocated in governing marine resources and pollution. Since many environmental problems are fraught with system uncertainties and incomplete information about the system characteristics, the Principle tends to be equally applicable, especially in problems like greenhouse effects. The increasing role of PP suggests that it is ripening into a norm of international law; some of the key elements of a legal definition rely upon the following: (a) a threshold of perceived threat against which advance action would be deemed justifiable; and (b) a burden of proof on the activity contributor or entrepreneur to show that a proposed action will not cause actual harm (Cameron and Abouchar, 1991). The PP implies current commitment of resources in order to safeguard against the likelihood of future occurrence of adverse outcomes of certain activities. This approach is implicitly seeking tradeoffs in the interests of the present with those of the future, and thus depends on implicitly assumed time discounting and future resource valuation (Rao, 1999). These factors are usually not examined in the current practices in the application of the Principle. This is because the role of the PP is largely confined so far to provide guidance to policy judgment and providing benefit of doubt in favor of the environmental resources.

The PP is equivalent to 'risk averse' behaviour in cases that involve irreversibilities or extremely high costs in socioeconomic or biogeophysical or other terms. When sought to be applied in general situations not necessarily involving these features, the PP could lead to a caution that may be attained at the expense of substantial potential gains. It would be similar to imposing a ban on any driving of an automobile since there is a positive probability of an accident in that activity. However, the risk averse nature of the Principle is very relevant if scientific knowledge is too limited to quantify uncertainty and thus cannot establish probability distributions of potential outcomes. The WTO framework provides a limited application of the PP under the SPS Agreement (regarding Sanitary and Phytosanitary Measures), but fails to utilize some of the requisite approaches in other agreements like GATT 1994 and TBT.

Another important principle more widely accepted, but limited in its actual practice is that of a prescription for internalization of environmental externalities. A greater integration of trade and environmental objective to their mutual reinforcement is likely with a judicious application of the imperatives of such guiding principles. This is the Polluter Pays Principle (PPP), discussed in the next section.

7.7   The Polluter Pays Principle

The 1972 OECD Guiding Principle concerning Polluter Pays Principle (PPP) (OECD, 1972) proposed a cost allocation principle that seeks to internalize costs of various transactions. It stated: 'the polluter should bear the expenses of...measures decided by the public authorities to ensure that the environment is in an acceptable state'. If properly applied, an application of the PPP tends to minimize trade distortions caused by environmental and other inherent subsidies by closing the gap between market prices and social costs of production (which include costs of negative externalities).

The PPP states: when, for example, the environmental costs of deterioration are not adequately reflected in the (market or administered) price system, the scarcity value of resources is not depicted; such costs should be incorporated in the assessment of goods and services.

As stated in OECD (1995), the PPP is concerned with 'who' should pay for environmental protection, and not with 'how much' should be paid. The 1972 Guiding Principle PPP also considered the issue whether countries should have the right to use border adjustment, duties or tariffs to offset the international differences in environmental costs, and rejected it for fear of potential discriminatory trade practices. The GATT as well as the OECD (1972) do not authorize the enforcement of the PPP with regard to imports to offset the costs of pollution prevention/control measures affecting the local and/or global environment. The likelihood that trade policies could affect environmental objectives of sustainable development was not considered when the 1972 PPP was drawn up. It was a very good principle for that time point, and now needs revision to its applicable forms. Natural resource conditions and environmental features specific to different countries also comprise of varying natural environmental assimilative or renewal capability of the corresponding sinks and sources.

Weak Polluter Pays Principle (WPPP)

This is a diluted version of the original PPP. This is built on the premise that when pricing on the basis of full costs (production costs plus environmental costs) generates its own set of negative externalities, the latter effects may be used to partially offset the original cost elements. This allows for a limited degree of 'subsidies' relative to the full cost, and recognizes both first order and second order effects of price interventions to correct a market failure. An element of economic feedback (both negative and positive) mechanism is utilized here in assessing the consequential effects as well. In the latter category one could also visualize the costs of forcing economic entities out of business as a result of some of the strict and narrow application of 'environmental principles'. Thus, this characterization of the WPPP is an aspect of integrated and pragmatic decision making. Thus, it promises greater role in recognizing trade–environment interdependencies and complementarities.

As an application, payments to farmers to adopt pollution minimizing practices of agricultural input use or other farm production methods, amounting one form or the other of subsidy or provision of incentives to adopt environmental quality-enhancing practices constitutes an example of WPPP (see Tobey and Smets, 1996, for a detailed application in the agriculture sector).

In a submission by Japan to the WTO/CTE, it was argued: Agriculture is multi-functional, such as preventing soil erosion, landslides and flooding, conserving water resources, preserving biodiversity, maintaining the landscape and providing recreational space, through sustainable agricultural activities in addition to its primary role of producing food and fibre and providing employment. Multi-functionality, which is inherently expressed specific to the natural conditions of each region, cannot be transferred and adapted to other regions, nor be substituted by the expression found in another region. Moreover, it should be noted that landscape and biodiversity, once they are lost, are impossible or very difficult to recover. When the environmental benefits from the multi-functionality of agriculture cannot be internalized, as is often the case, some policy measures are necessary to complement the market mechanism. Agricultural Ministers acknowledged in the Communique of the OECD Meeting of the Committee for Agriculture that: 'there can be a role for policy where there is an absence of effective markets for such public goods, where all costs and benefits are not internalized'. As agricultural trade liberalization itself cannot cope with so-called 'market failures', active government interventions are necessary in order to maintain and enhance sustainable agricultural activities which express multi-functionality in each region.

The above description, justification or otherwise apart, point to the active interaction of policies regarding trade liberalization, internalization of environmental costs and benefits, and the role of government interventions in managed trade. The role of the WPPP is even more apparent in many such situations.

Strong Polluter Pays Principle (SPPP)

This is an extended version of the PPP. This warrants that the costs of damage to the environment should be calculated on the basis of the life-cycle of the products and sought to be paid up front by the producer/supplier (in the chain of provision, somewhat similar to the methods used in a few countries that use multi-stage value-added-tax (VAT)). As deliberated in Section 7.3 above, the method of environmental cost internalization on the basis of SPPP may neither be feasible nor desirable in all production and trade scenarios.

The GATT rules tend to support externalization of environmental costs as legitimate source of comparative (and possibly competitive) advantage, with an obsolete prescription that the parties cannot 'discriminate between like products on the basis of the method of production'. The sooner the concept of externalities sinks into the participants of the WTO Council, the world will be saved of potential environmental catastrophes and other environmental costs.

The GATT rules need to be amended to allow countries to object to import of goods using production processes that create transnational spillovers, stated Bhagwati (1993) who also suggested that: any environmental safeguards should consist of an obligation to demonstrate that the ETM is designed to promote an environmental 'solution that is both efficient and equitable' among the member countries.

When multilateral trade policies are required to recognize internalization of environmental costs and multilaterally agreed principles of PPMs, environmental and trade objectives tend bear greater coherence. The 'efficiency-enhancing' role (Bagwell and Staiger, 1999) of the GATT principles of non-discrimination and reciprocity will continue to hold even when these environmental criteria are incorporated, provided the latter are done on a multilateral basis. This becomes a possibility as long as the externalities of environmental criteria are not allowed to lead to discrimination and all such effects are captured by the global price mechanisms. The secretariats of the MEAs and the WTO should coordinate to achieve this delicate balancing of trade and environmental objectives to the mutual reinforcement of each other.

The economic principles such as the PPP are usually inadequate to govern socially and environmentally desirable and sustainable trade patterns. A set of broader foundations including the rules governing tradable items are necessary before economic principles can be properly utilized. The vexed issue of trade in known harmful chemicals is an important sector which merits greater attention. The next section reflects on the lack of a multilateral trade policy regarding exports in such dangerous products (see also Box 5.2).

7.8   Domestically Prohibited Goods for exports

The issue of exports of DPG was discussed as early as in 1982 at the GATT Ministerial Meeting. Seven years later a Working Group on Export of DPG was constituted in 1989. A Report called 'Draft Decision on Products Banned or Severely Restricted in the Domestic Market' was submitted in 1991 by this Group to the GATT. However, this Report was not processed further and the Working Group never met again. At the end of the Uruguay Round deliberations, it was agreed in the 1994 Marrakesh Ministerial Decision on Trade and Environment to incorporate these issues into the work programme of the CTE of the WTO. Five years later in 1999, it was reported that the CTE 'will continue to examine what contribution WTO could make in this area, bearing in mind the need not to duplicate work of other specialized agencies' (WTO, 1999).

The voluntary code of ethics was developed in response to decision 16/35 of the Governing Council of the United Nations Environment Programme (UNEP) of May 1991 entitled 'Toxic chemicals', and to Agenda 21, in particular its chapter 19 and chapter 30 (paragraph 1) on environmentally sound management of toxic chemicals which was adopted by the United Nations Conference on Environment and Development at Rio de Janeiro in June 1992 and endorsed by resolution 47/190 of the United Nations General Assembly in December 1992. The code is a complement to the amended London Guidelines for the Exchange of Information on

Chemicals in International Trade) which address Governments and the scope of the code is broader than that of the amended London Guidelines.

The Code is the outcome of a series of UNEP consultative meetings for private sector parties convened between May 1992 and April 1994 in response to decision 16/35 of the Governing Council of UNEP and chapter 19 of Agenda 21. The First Session of the Intergovernmental Forum on Chemical Safety (Stockholm, April 1994), in its Priorities for Action, recommended that, as a particular priority, the Code should be applied widely by industry in all countries without delay, and encouraged as promoted by the Code the circulation of safety data sheets for all dangerous chemicals in international trade. The second session of the Commission on Sustainable Development of the United Nations (New York, May 1994) endorsed those recommendations.

The nineteenth special session of the United Nations General Assembly (New York, June 1997), in the Programme for the Further Implementation of Agenda 21 adopted by its resolution S/19-2, recognized the finalization and application of the Code as one of the major achievements in the implementation of Agenda 21 with respect to environmentally sound management of chemicals.

7.9 Concluding observations

The importance of integrated approaches to trade and the environment is better appreciated when the phenomena of global financial and economic integration are understood in terms of their implications on the local, regional and global environment. The fundamental principles, the PP, and PPP, do not apply only to the trade–environment nexus. If these aspects are not reflected in one form or the other in economic and other activities (such as environmental liability of global lenders) we are unlikely to address the problems in their full dimensions.

Based on data for 1995 (IMF, 1996), the total debt of fuel exporting countries consists of about one-eighth of the global debt of all developing countries; total exports of goods and services account for a little less than the above share in the corresponding group. However, primary mineral exports tend to have much lower share (about 2.2 per cent) of export value, and higher share of debt (about

5.1 per cent) among developing countries. It is unlikely that the free market works as freely when the exporters are constrained by debts and repayment obligations in tight schedules. The ability of some of the exporters to adhere the PPP or other criteria remains is severely constrained. Multilateral trade policies under the WTO regime should address these issues on a much broader and clear footing than conducted so far.

The trade regimes governing (rather ungoverning) DPG is a continuing global environment-trade problem. The failure of the WT/CTE to come up specific multilateral trade rules could necessitate a separate MEA – somewhat similar to the Montreal Protocol to phase out the CFCs. This does not portend a great role for the WTO system. The lesser the WTO framework and its subsidiary organization appreciate the need for effective integration of trade and environmental issues, the greater the potential for this unique institution to be marginalized and undermined over a period of time. Hence the task lies ahead for the WTO to come up with greater timeliness and focus to the relevant principles and practices.

References

Bagwell, K. and R. Staiger (1999) 'An economic theory of GATT', *American Economic Review*, 89.1, 215–48.

Beghin, J., D. Roland-Holst and D. van der Mensbrugghe (1994) 'A survey of the trade and environment nexus – global dimensions', *OECD Economic Studies* #23, pp. 167–92, Paris: OECD Secretariat.

Bhagwati, J. (1993) 'Trade and the environment – the false conflict?', in Zaelke *et al.* (1993), pp. 159–90.

Cameron, J. and J. Abouchar (1991) 'The precautionary principle – a fundamental principle of law and policy for the protection of the global environment', *Boston College International and Comparative Law Review*, 14, 1–27.

Elliot, G. (1994) 'Internalization of Environmental Costs and Implications for the Trading System', GATT Symposium on Trade, Environment and Sustainable Development, Geneva: GATT.

Francois, J. and K. Reinert (eds) (1994) *Applied Trade Policy Modeling*, Cambridge: Cambridge University Press.

IMF (1996) *World Economic Outlook*, Washington, DC: IMF.

Kapp, K. W. (1963) 'Social costs of business enterprise', in Massarrat (1997), p. 30.

Low, P. (1993) *Trading Free – the GATT and US Trade Policy*, New York: The Twentieth Century Fund Press.

Khalid, A. R. and J. B. Braden (1993) 'Welfare effects of environmental regulation in an open economy – the case of Malaysian palm oil', *Journal of Agricultural Economics*, 44.1, 25–37.

Kox, H. (1995) 'LDC Primary Exports and the Polluter-Pays-Principle: a Case for International Policy Coordination', UNCTAD Expert Group Meeting on Internalization of Environmental Costs and Resource Values, Geneva: UNCTAD.

Lee, H. and D. Roland-Holst (1994) 'International trade and transfer of environmental costs and benefits', in Francois, J and K. Reinert (1994).

Massarrat, M. (1997) 'Sustainability through cost internalization – theoretical rudiments for the analysis and reform of global structures', *Ecological Economics*, 22.1, 29–39.

OECD (1972) *The Polluter Pays Principle*, Paris: OECD Secretariat.

OECD (1994) *The Environmental Effects of Trade*, Paris: OECD Secretariat, pp. 8–12.

OECD (1995) *Environmental Principles and Concepts*, OECD Document OCED/GD(95)124, Paris: OECD Secretariat.

Porter, G. and J. W. Brown (1991) *Global Environmental Politics*, Boulder: Westview Press.

Rao, P. K. (1999) *Sustainable Development – Economics and Policy*, Oxford & Boston: Blackwell Publishers.

Salzman, J. (1998) 'Informing the green consumer – the debate over the use and abuse of environmental labels', *Journal of Industrial Ecology*, 1.2, 11–21.

Siebert, H. (1996) 'Trade policy and environmental protection', *The World Economy* (Global Trade Policy Issue), 183–94.

Tobey, J. and H. Smets (1996) 'The Polluter Pays Principle in the context of agriculture and the environment', *The World Economy*, 19.1, 63–88.

Tuchband, M. (1995) 'The systemic environmental externalities of free trade – a call for wiser trade decision making', *Georgetown Law Journal*, 83.5, 2099–118.

UNCTAD (1993) *Trends in the Field of Trade and Environment in the Framework of International Cooperation*, UNCTAD Report TD/B/40(1)/6, Geneva: UNCTAD.

UNCTAD (1994) *The Internalization of Environmental Costs and Resource Values – a Conceptual Study*, UNCTAD/COM/27, Geneva: UNCTAD.

UNCTAD (1995) *Sustainable Development and the Possibilities for the reflection of Environmental Costs in Prices*, Geneva : UNCTAD Report # TD/B/CN.1/29.

UNCTAD (1997) *Trade and Environment – Concrete Progress Achieved and Some Outstanding Issues*, Geneva : UNCTAD Report.

UNEP (1996) *UNEP – World Bank Workshop on the Environmental Impacts of Structural Adjustment Programs, March 1995*, UNEP Publication Economic Series #18, Geneva: UNEP.

Vainio, M. (1998) 'The effect of unclear property rights on environmental degradation and increase in poverty', UNCTAD Discussion Paper #130, Geneva: UNCTAD.

Walter, I. and J. Loudon (1986) 'Environmental costs and patterns of north–south trade', paper prepared for the World Commission on Environment and Development, New York: United Nations.

WTO (1999) *Background Document for the WTO High Level Symposium on Trade and Environment* (15–16 March 1999) Geneva: WTO Secretariat.

Zaelke, D., R. Hausman and P. Orbuch (eds) (1993) *Trade and the Environment*, Washington, DC: Island Press.

8
New Role for the WTO

8.1   Revising the WTO articles for environmental provisions

Much of the approach and specifications in the WTO charter regarding the environment is at best one of conciliation and concession rendering rather than one of recognition of direct and indirect linkages between economic and environmental features, their dynamic interdependence and the design of policies with less than impatient exploitation of resources of the planet. Despite an eloquent statement in the preamble to the WTO Article of Agreement, little else seems to contribute to that proposed foundation. No doubt, the WTO is not to be deemed as an environmental protection agency and that it should maintain its focus on the governance of the multilateral trading system as per its charter. This assertion also specifies a duty to act: the specified tasks include, *inter alia*, coherence with within and across various articles of the WTO charter. The WTO working rules cannot ignore the foundation in the preamble. Such a mechanism is not expected to enhance the functioning of the institution nor its contribution to the stated objectives.

The preamble relating to requirements of protection of the environment and natural resources, compliance with sustainable development in the WTO Articles did not get reflected in the new GATT 1994; its interpretation for application of trade and environmental principles continues to fail to recognize the role of environment in most trade disputes. Since the new GATS is more like a twin entity of the GATT, the same terminology and interpretation of various clauses tends to hold. The Agreement on Sanitary and Phytosanitary Measures deals only with short-term measures and does not recognize some of the adverse health effects of harmful chemicals being exported. The following provide some of the important requisite changes in the Articles of the main agreements. Undoubtedly, these are neither exhaustive nor are aimed at absolute environmental protection for its own sake. Rather, these are expected to suggest a reconciliation of trade and environmental objectives with a clear provision for the environmental safeguard, including in those agreements where is no mention of the word 'environment'.

Some of the countries/blocs such as Norway, Switzerland and the European Commission proposed to strengthen integration of the trade and environmental policies during the December 1999 Seattle Ministerial Conference of the WTO. The proposals sought specifications of the exact relationships between the MEAs and the multilateral trading system governed by the WTO.

GATT 1994 and GATS

The GATT 1994 remains the focal point of trade disputes and their environmental 'exceptions'. The 'new' GATT is not any new as far as its treatment of issues relating to the environment are concerned. As long as the word 'environment' remains a strange term for the GATT and its exceptions under Article XX, it is futile to argue whether or not the DSU and the DSB will be able to integrate environmental considerations into trade polices and resolution of trade disputes.

Article XX(b) was proposed for amendment by the Austrian proposal during the Uruguay Round negotiations, to read: 'necessary to protect the environment, human, animal or plant life or health' (GATT, 1990). It was considered 'too late to start working on it in the Negotiating Group. No effect was given to this proposal'. It was after four years of the proposal that the Uruguay Round ended, and it is about a decade since that time that no meaningful action has been taken yet on this clause. The proposal was essentially to include the word 'environment' for the first time in the GATT text after about four decades of its provincial existence, but that was not accomplished either.

Based on later negotiations, GATS, now part of the WTO Agreement has GATS Article XIV which is analogous to GATT Article XX.

It is meaningful to revise both with relevant reforms:

1 Include 'environment', in addition to the usual set of items in GATT XX(b) and GATS XIV(b); and,
2 Replace 'exhaustible resources' with 'nonrenewable resources' in order to include a larger set of resources and under less severe extinction conditions.

There exists little congruity in the narration of different clauses under Article XX, especially in relation to XX(b) and XX(g): the operative part of the former starts with 'necessary', and that of the latter focuses emphasis on the 'primarily aimed at' requirement. The former seeks a cost-minimizing sense (seek the lowest cost alternative), and the latter tends to be satisfied with less rigorous application (as long as other main clauses of the GATT are complied with). It is useful to revise the expression to clearly state that the exercise of the exceptions provisions should be based on a meaningful analysis of cost-effective alternative, where the elements of costs could include admissible realistic transaction costs as well. The role of preserving 'exhaustible' resources must be recognized in terms of that of 'non-renewable' resources. This amendment is expected to protect the interests of (a) future generations with greater significance, and (b) enhance the value of fast depleting resources, including the potential compensation to the providers of most primary commodities. Let us also note that many of these primary commodity suppliers in international trade are relatively poor countries and that these tend to be adversely affected by the existing unsustainable terms of trade patterns. In other words, the suggested amendment has regional and intertemporal dimensions justifiable on efficiency and equity grounds.

It is desirable to amend the GATT Article XX to resolve some of the potential conflicts between various provisions of GATT and some of the MEAs. This amendment could be similar to some the exceptions under GATT Article XX(h) for approved trade measures implemented in response to international commodity agreements. This would facilitate member countries to adopt ETMs mandated under MEAs without being GATT-inconsistent.

SPS and TBT agreements

The Agreement on SPS recognizes only what meets the eye: health concerns of life forms if these are direct and indirect; it does not seem to matter if the human health problems are an event of the future, however certain such an adverse effect might be. Adverse health effects of ozone depletion or greenhouse gas concentrations are not even remotely addressed in any of the agreements under the WTO charter. These aspects were supposedly covered (or to be covered, whenever) under one or more of the MEAs. To the extent that some of these global environmental problems tend to be exacerbated by various trade mechanisms, the WTO has an obligation consistent with its preamble to the charter. The environmental policy measures to mitigate these rather irreversible problems should also be equal concern in the multilateral trade policies and their promulgation under the trade mechanisms. Let us recognize that the global commons problems have their origins in microeconomic decisions, governed by macroeconomic policies including those governing global trade.

Accordingly, it is important to formulate policies within the WTO framework to mitigate adverse environmental effects of endorsing the following: processes and production, trade, and consumption, of items of that aggrevate global environmental problems only to the extent 'necessary'. Just as the interpretation of 'necessary' clause was provided in attempts to resolve trade disputes under the GATT Article XX, it is useful to incorporate the provision that process and production methods be environmentally efficient, and that this is to be interpreted in terms of adoption of process and production measures causing environmental damage at levels 'necessary' in the sense of minimizing total costs over time (including transaction costs, and recognizing a reasonable discount on time for the valuation of future costs). These observations are equally relevant in suggesting modifications to the Agreement on TBT as well. The TBT Agreement states the measures under the Agreement shall 'not be more trade-restrictive than necessary'. Analogous reasoning should also simultaneously hold for ensuring that the eligible measures are 'not more environment-damaging than necessary'. The reference here is primarily to the global environmental features, and the protection of the local and regional environment is expected to be sought under a number of other instruments of policy, including the PPP.

The SPS Agreement recognizes the right of governments to take precautionary provisional measures when scientific evidence is lacking, while obtaining further information to support or revise a proposed non-discriminatory trade measure covered under the Agreement. When a trade dispute arises over any such measures, the Agreement provides for the role of a scientific expert panel for resolution of the issues. While these constitute improvements over many similar Agreements, the insistence on the use of international organizational standards may need further explanation in order that the stated guidelines can be observed in practice in most developing countries as well. Besides, the provision for application of the PP is severely restrictive; a broader application, with the provision that the application be made transparent and reviewed periodically by a scientific panel, is likely to be more productive. Potential misuse of such wide ranging exceptions creates a few problems, but a recorded (as per the scientific analyses) systematic pattern of such practice of any member country can provide a reasonable disincentive against misuse. Additional disincentives could also be devised when permitting the application of the PP. The main aspect to be recognized here is that societies in member countries should have a choice – within reasonable limits – to exercise sovereign rights to select their own specifications of types and magnitudes of risk-taking in environmental-and-public health governance. Since the PP does not lead to any uniquely or precisely specified levels of acceptable risks, individuals members need to exercise their own informed judgments, utilizing scientific information. This is typically an exercise in decision-making under uncertainty with incomplete information. The issue that pertains to the WTO is to ensure that arbitrary discrimination of trade activities will not be product of the application of the PP. Additional guidelines need to be worked out in this regard.

Agreement on subsidies and countervailing measures

The provisions regarding environmental subsidies in the 'non-actionable category' remain rather restrictive. Since R&D is the key to technical progress and control of adverse effects of environmental pollution, and also since compliance with various MEAs depends heavily on R&D, greater incentives and broader provisions are required in this category. A continuing 'non-actionable' exemption rather than a one-time upgradation provision is required for adoption to new environmental regulations. Besides, there is little clarity in the specifications regarding process R&D and product R&D, even after specifying so-called basic R&D. The economics of R&D clarifies that both process R&D and product R&D are largely complementary. It is also well known that there are several positive externalities of R&D which benefit much of the world in various development aspects. Thus, it may not be meaningful to be very restrictive in providing exceptions regarding subsidies. However, more detailed guidelines will be required in this regard.

8.2   Trade policy review mechanism reform

The current practices under the TPRM are not well suited to integrate the environmental concerns of trade and trade concerns of the environment. This aspect is not specified in the provisions of the TPRM either. It can also be stated that the current practice does not conform to the specifications either. Lack of direction, structure and framework is noticeable where the participants comment on a wide range of issues from predominantly political perspectives (see Quereshi, 1995 for a set of examples in this regard). Some of the recent reviews are dominated by the viewpoints of the other global institutions like the IMF who seem to steer the processes from backseat driving. This approach distorts the objectives of the reviews and do not lead to constructive approaches to the relevant issues.

The regular Annual Report of the Director General WTO regarding TPRB does not deal with trade and environment interface issues at all. This situation requires changes, preceded by changes in the TPRM to reflect environmental concerns of trade policies. The TPRM should provide for a careful review of the trade and environment integration possibilities, as also those of trade and development. An environmental accounting exercise (on the lines of the UN system of environmental accounts – for details see Rao, 1999) will complement the efforts to ensure that trade and other activities are in line with the requirements of sustainable development. Only then a comprehensive and operationally relevant exercise of the review can be meaningfully devised and followed upon for implementation. It is useful to note that the March 1999 WTO High Level Symposium on Trade and Environment broadly supported the idea of conducting environmental reviews of trade agreements. A similar reasoning should be used for conducting environmental reviews during member-specific trade policy reviews and their interface with the member's obligations under various MEAs and multilateral trade agreements under the WTO framework.

8.3   Multilateral environmental agreements and the WTO

The provisions involving the interdependencies of trade and environment and those of trade and development are rather conspicuous by their extremely marginal roles in the WTO framework. The stand maintained by some that environmental issues are best addressed by MEAs is only partially meaningful. Besides being marginalized by seeking the MEAs to address these issues, including use of ETMs, the WTO will adversely affect the potential cost-effective solutions that exist when trade and environmental measures are devised in an integrated framework. Unless separate treatment is justified by institutional and transaction cost considerations, there is no reason why the WTO should refrain from active policy design for multilateral trade regimes that offer sustainable trade and development patterns for all countries.

Several MEAs rely on the use of PPMs to achieve environmental objectives. For example the UN Framework Convention on Climate Change, which led to the 1997 Kyoto Protocol mandates several countries, most of whom are members of the WTO, to control their greenhouse gas emissions with appropriate production regulations. These have implications on relative competitiveness and on multilateral trade. The WTO framework should coordinate with the MEA secretariats to ensure that implementation of such mandates does not create any trade disputes in the subsequent phases. The problem of 'greenhouse gas leakages' can arise when unaccounted emissions of greenhouse gases accumulate in a 'free-rider' regime where a country is neither a member of the WTO nor has obligations under the MEA. Advance specification of incentives for such countries to comply for in order to become eligible to seek any WTO membership will be necessary. Similarly, disincentives like trade restrictions may be desirable to motivate the country or group of countries to account for greenhouse gas leakages beyond reasonable limits.

In regard to known harmful chemicals, the WTO should play a more positively active role to bring about a meaningful multilateral trade agreement so as to safeguard the immediate interests of affected developing societies, and also to protect the exporter communities from imports based on the same harmful chemical exports. Rather than expecting MEAs to handle more trade measures to control production, trade and use of these chemicals, the WTO should coordinate such efforts. It should also be feasible to seek phase out of these items and provide concessional terms for the developing countries to adopt new alternatives that are environmentally sensible. Besides, the WTO should participate in the implementation of the 1998 Rotterdam Convention on Prior Informed Consent for Hazardous Chemicals Trade.

8.4   WTO and sustainable development

The WTO can make an important contribution to the work of 'optimizing the economic system' and ensure sustainable trade, rather than freer trade based on relatively short term considerations and heavy reliance on mercantalism. The WTO's specific contribution would be to ensure that future trade negotiations give priority to those trade reforms that either directly or indirectly can facilitate better environmental outcomes. These reforms must include action to eliminate or reduce or ensure adequate discipline is imposed on trade-distorting subsidies.

The WTO should adopt the guiding working principle of ensuring the contestability of international markets to achieve efficiency of trade activities. When the sustainable development principles constitute the guiding philosophy of trade, the working principles that lead to this desired configuration must be devised, consistent with the WTO charter and objectives of multilateral trade system.

Some of the main working principles relevant for the WTO framework are:

1   The PP
2   The PPP
3   New Trade Regimes for the Control of Harmful Chemicals
4   Adoption of Environmental Accounting in each member country
5   TPRM to include environmental review.

The WTO framework should remain rule-oriented with greater transparency. This requires that the WTO General Council should devise appropriate provisions for environmental considerations;

DSB or dispute resolution panels cannot substitute their judgment for the WTO. The recommendations of various committees working under the WTO charter should be prescribed to provide as much of transparent and clear guidelines as possible in a time bound manner on any specific issue; bureaucratease should be replaced by results-oriented approach.

On the issue of harmonization of environmental standards, an 'optimal degree of harmonization' is to be explored in harmonizing standards. Relative movements of standards and the role and limitations of least common denominator need to be assessed before prescribing standards on a universal basis. Thus far there exists little understanding of a comprehensive benefit–cost analysis in this regard. In respect of possible adoption in the developing countries, the WTO should attempt to mediate more meaningful trade policies and special concessions to severely indebted developing countries, and also coordinate relevant technical assistance from other participating international organizations.

Among significant undocumented non-tariff barriers to trade are corruption factors affecting some of the trade activities in some of the member countries. Corruption acts as a barrier to international trade – it blocks trade by acting as a surcharge on goods and services, and creates de facto monopolies controlled by the state machinery; it is thus a 'non-tariff barrier' to trade (Nichols, 1997). The preamble to the WTO charter promises optimal use of resources and sustainable development. The WTO might consider resolving disputes related to failure of one or more members in fulfilling their obligations under the agreements.

The WTO should cooperate with the Global Environmental Facility (operating under the Montreal Protocol) to integrate some of the environmental problems affecting the global commons, and delineate the role of environmental trade measures in achieving stated environmental objectives with trade practices. Since the emerging global environmental problems are of global commons category, it is unrealistic to believe that the MEAs are adequate to address these issues (with or without environmental trade measures). After all, these problems of the global commons have their roots at the micro-level economic activities. Global trade is not necessarily the main contributor to these problems, but it is not an uncorrelated contributor either. The role of the WTO should be perceived as that of ensuring sustainable trade with a multilateral system, and the preamble of the WTO charter is a good beginning for attaining the main objective, provided the preamble is seen not as an end but only as a beginning.

Third Ministerial Conference: Seattle

The WTO Secretariat as well as several member countries prepared a number of background notes and special studies for the Seattle Conference held during 30 November to 3 December 1999. In a study on trade and environmental linkages coordinated by the WTO Secretariat, Nordstrom and Vaughan (1999), the role of the environment was largely sought to be handled via multilateral environmental institutions and mechanisms rather than under the framework of the WTO charter. It was suggested that much of the focus on the WTO was founded on the provisions of 'trade sanctions' under the WTO mechanism (and lack of such comparable actionable instruments under various MEAs). Among important positive roles suggested for the WTO were: 'facilitating the diffusion of environmentally friendly technology', and effecting a reduction in the 'cost of investing in clean production technologies and environmental management systems'.

While moving in the right direction on some of the issues, the Nordstrom and Vaughan (1999) study fell short of arguing for a case under the TRIPS and SCM Agreements to facilitate a transparent provision for promoting environmentally efficient technologies under the multilateral trade mechanisms. Role of technology subsidies for adoption and transfer could be improvised, somewhat on the lines of the provisions of 'non actionable' subsidies for R&D. The SCM Agreement allows the following, and needs a more comprehensive provision to include incentives for promoting environmentally efficient technologies (beyond an acceptable minimum improvement level: SCM Agreement Provision on 'Assistance to adapt existing facilities to new environmental requirements': 'Such assistance must be on a one-time basis, be limited to 20 per cent of adaptation costs, and be available to all firms which can adopt the new equipment and processes.'

Let us note that pollution prevention and pollution control are two broad categories of environmental measures, and these are complementary to each other both in the static and in the dynamic context. As argued by Gollin (1991), much of the conventional environmental regulation maintained emphasis at 'restricting the use of harmful technology than at promoting innovative beneficial technology.' It is feasible that TRIPS can also be modified for the provision of greater incentives for relevant innovations. In addition, a revolving capital fund may also be created to enable concessional lending to promote relevant environmental technology trade, possibly in conjunction with the World Bank and regional development banks.

On the eve of the Third Ministerial Conference in Seattle, the WTO Secretariat circulated a brief note entitled 'What's at stake?' It was stated that one of the reasons why environmental issues remain 'controversial' in the WTO is that 'some developing countries fear that environmental measures may be used, deliberately or not, to create barriers to their exports'. It is desirable that such fears are allayed with the systematic development of transparent multilateral trade and environmental policies and operational rules for the mutual benefit of all members, as envisioned in the preamble for the WTO charter.

The trade ministers of 135 countries held protracted deliberations in Seattle. These did not lead to any significant developments in the global policies of multilateral trade and environment. The Seattle trade talks have now been recognized for what they failed to achieve rather than what they did achieve to fulfill the stated objectives of the WTO charter or its reform. Perhaps the conference was an illustrative case of how not to deliberate for a potential consensus. The Seattle talks were originally anticipated to launch a new round of trade negotiations, possibly to be named either the 'Seattle Round' or the 'Millennium Round'. The next round of trade negotiations starts later in the year 2000, based on another set of consultations of the WTO members in Geneva. The Seattle talks could not generate even a common agenda for further deliberations. In fact, the conference was scheduled (in accordance with a previous year's resolution in Geneva), with almost the entire proposed agenda stated in brackets, that was to say there was little *ab initio* agreement among members regarding proposed agenda for negotiations. To add to the complexity, most developing countries (constituting about two-thirds of the WTO membership) expressed distrust in the manner in which the forum was conducted, including the formation of 'informal working groups'. An unprecedented show of protests by various environmental groups and labour unions added to the complexity of consensus building on any major issue. The developing countries did not evince inclination to support some of the proposals on environmental and labour standards for fear that these would lead to disguised trade protectionism in the developed countries.

The WTO charter and its preamble provide guidance to derive pragmatic multilateral trade policies consistent with environmental preservation and sustainable development. The member countries as well as the WTO Secretariat need to undertake necessary analyses and devise mechanisms to ensure a fair and equitable global trading regime. Such efforts can lead to confidence building for all the members and enable a reasonable consensus. It is also useful to realize that much of the debate on trade and development in the WTO framework (and under the GATT 1947 framework earlier) has been narrowly focused in terms of trade and developing countries, or in terms of developed and developing countries. This was also evident from the WTO High Level Symposium on Trade and Development held in Geneva during March 1999. Most of the modern concepts and approaches for development and sustainable development are still missing in their integration with multilateral trade mechanisms under the WTO charter. Several ingredients relevant for strengthening the framework are enunciated in this book within the context of trade–environment sustainable development.

References

GATT (1990) GATT Document MTN.GNG/NG7/W/75 (1 Nov. 1990), in WTO (1999).

Gollin, M. A. (1991) 'Using intellectual property to improve environmental protection', *Harvard Journal of Law and Technology*, 4, 193–235.

Nichols, P. M. (1997) 'Outlawing transnational bribery through the World Trade Organization', *Law and Policy in International Business*, 28.2, 305–81.

Nordstrom, H. and S. Vaughan (1999) *Trade and Environment*, WTO special study #4, Geneva: WTO Secretariat.

Quereshi, A. H. (1995) 'Some lessons from developing countries' trade policy reviews in the GATT framework – an enforcement perspective', *The World Economy*, 18.3, 489–504.

Rao, P. K. (1999) *Sustainable Development – Economics and Policy*, Oxford & Boston: Blackwell Publishers.

WTO (1999) *Background Document WTO High Level Symposium on Trade and Environment* (15–16 March 1999), Geneva: WTO Secretariat.

9
Towards a Better Future

9.1 Multilateral trade, environment and development

If the sole objective of a multilateral trade regime is to ensure smooth and predictable functioning of the global trade activities, with little regard for the interlinks with environmental sustainability, the trade system will not continue to remain smooth nor predictable. The role of information and institutions is very significant in mitigating potential problems.

The operational objectives of multilateral trading system need to be formulated in terms of market institutions and their governance. The issue of governance must be better appreciated in terms of the implications on transaction costs, comprising mainly the costs of undertaking trade transactions as well as the costs of devising institutional arrangements, monitoring and enforcement. The role of welfare enhancing competition policies on a fair and non-discriminatory basis remains the foundation for all approaches toward multilateral trade governance. The natural forces operating on the market institutions constitute the fundamental dynamics, and these can only be partially influenced by the rules of governance. At the crux of the matter remains the issue of 'contestability of markets' (see Baumol *et al.* (1982) for a detailed account of the features). This feature may be characterized in terms of the prevalence of market conditions of the competitive processes wherein the anti-competitive forces due to government or private actions are eliminated so as to ensure unimpeded market access (see also Zampetti and Sauve, 1996).

The primary role of the WTO may be viewed in terms of its effective contribution to the creation and sustenance of international contestability of markets. This characterization remains invariant under its obligations to ensure sustainable trading patterns and contribute sustainable development processes with an integration of trade–environment–development issues. The WTO rules should attempt to protect the competitiveness of the global markets, and not necessarily the competitors. In other words, protecting the competitors does not usually amount to ensuring the workability of contestable competitive markets. The main reason for this process is built on the dynamic effects and externalities of such protective devices, direct or indirect. As long as the institutions are appropriately devised and sustained to govern the multilateral trading systems, the operative rules apply at each time instant to ensure that any deviations built over time (possibly as a result of one or more specific interventions) are brought in line for conformity.

Another important feature of international trade, not fully recognized in the WTO framework relates to the role of implicit trade barriers posed by large conglomerates and cartels with monopolistic powers to alter the competitiveness of various markets. Some of the rules under the WTO address these concerns: Article VIII of the GATS stipulates that member countries should ensure that their monopolistic enterprises do not exercise undue market influence. Similarly, Section 8 of the TRIPS Agreement provides measures against anti-competitive practices in contractual licenses (see also Immenga, 1998). The role of an international antitrust code remains an issue for further consideration within the WTO framework. This should be seen as an element of the need to ensure international contestability of markets.

The role of multilateral trade mechanisms should also be to address some of the issues of global economic inequities and their continuous deterioration. According to a recent assessment of the WTO, the global market factors, domestic political and economic crises in some of the developing countries, and the financial crises magnified by debt crises and exchange rate crises led to force down the share of primary products in the global trade of all goods to a share of 20 per cent (in 1998 prices). Let us recognize that the primary product producers are largely the countries in the least developed and developing categories. The result of global economic integration should not be allowed to deteriorate itself to contribute to some of these problems. The products share constituted the lowest level of the share since the end of World War II. Similarly, the value of traded goods declined for that year by two percent despite a growth in the volume of traded goods by about 3.5 per cent. Such factors, combined with ever increasing debt levels for developing countries, particularly debt servicing ratios to exports, do not suggest the formation of the most appropriate foundations for the functioning of an efficient global market. This market heavily constrained and not a 'free market' by any measure. A reasonably comprehensive analysis of the structure and functioning of the global trade systems would reveal the relative roles of various interlocking systems. It is useful to realize the contributory roles of constrained exports at largely pre-determined levels based on debt service requirements to fulfill continued credit access from global financial institutions for indebted countries, even under depressed 'market' prices and exchange rates. Attempts to devise and apply market principles without due recognition of the institutional factors can lead to misleading and unwise prescriptions for the vulnerable.

Let us recall an observation of Paul Samuelson (1962): free trade will not necessarily maximize the real income or consumption and utility possibilities of any one country; winners can potentially contribute to losers and gain from trade. An analogous extension holds good when we include the implications of trade gains for the environment. The operational implications is that gains from trade should be partly made available to offset the negative environmental impacts.

It is by no means less important that the multilateral trade system takes into account humane considerations, social dimensions of production and trade activities, the moral basis of trade expansion and protection of the vulnerable, be it the environment, or the poor working populations, the rights of humans and other living species. Indifference or insensitivity to some of these values tends to perpetuate a cycle of mutual exploitation and disrupt the natural cohesion of various systems (see the case of flower trade (Thrupp, 1995)). Material welfare is but a component of the determinants of human welfare. The role of trade fits into such a framework usually as a means of economic prosperity rather than pretend as an end product by itself. Accordingly, it is necessary that an integrated approach to trade issues is developed. It is apt to quote Theodore Roosevelt in this context. In his 1937 address to the US Congress President Roosevelt stated: 'Goods produced under conditions that do not meet a rudimentary standard of decency should be regarded as contraband and not allowed to pollute the channels of international commerce' (quoted in Kane, 1993). This prescription remains robustly valid today just as it was relevant several decades ago.

9.2   Institutions for sustainable trade and development

The WTO came into existence with a unique institutional configuration; it needs an effective institutional constellation, however. This is better achieved in both horizontal and vertical coordination with global institutions and stakeholders. The role of the WTO members remains the key to its effective contribution, but the role of its demarcated Committees and other working groups is of great importance as well. Given the initial base and platform with legal structure, the WTO is expected to contribute significantly toward fair and sustainable trade. The role of the regional trading arrangements remains important, but efforts should be made to ensure that these arrangements also synchronize with the objectives of the WTO framework.

The role of terms of trade in the broad patterns affecting the developing countries is unlikely to be a useful weakness for the global economies, given the features of their interdependence. The financial institutions should devise sustainable policies in various sectors and regions so that economic integration is achieved on a sustainable basis. It is useful to recognize that global economic integration does imply global environmental integration as well. Thus, there exists little rationale for devising economic or other policies which maintain neutrality to various environmental interdependencies across different societies, economies and time intervals. The degree of endogeneity of the state of the environment is directly affected by the 'quality' as well as quantity characteristics of trade and other transactions. Greater recognition of these issues and provision of rather comprehensive instruments to modulate the effects will be a globally useful activity for the WTO.

The need for greater use of environmental knowledge, possibly with an active cooperation with the Intergovernmental Panel on

Climate Change (IPCC), UNEP, FAO, and other internationally recognized scientific institutions will be a useful additional input into the decision making processes in the WTO systems. It is not entirely valid to maintain that any in-depth assessment would mean duplication of work of other institutions in the international network; none of these have a comparable charter and multilateral obligations to fulfil, and most do not have an active an active (or in principle active, as per the charter) membership role in formulating policies. This feature should be utilized as a strength for the institution and its effectiveness. Cooperation with institutions like the World Bank and IMF is desirable, but their mandate is entirely different. There are no 'client states' and shareholders for the WTO, but the business of the financial institutions is in terms of its clientele and shareholders. The issue of accountability is thus divergent as well. The WTO has a unique opportunity to contribute harmonious and sustainable development of the global systems. A recognition and contribution in this direction is a reasonable expectation for the WTO to fulfil.

References

Baumol, W. J., J. C. Panzar and R. D. Willig (1982) *Contestable Markets and the Theory of Industry Structure*, New York: Harcourt Brace Jovanovich.

Hope, E. and P. Maeleng (eds) (1998) *Competition and Trade Policies – Coherence or Conflict?*, New York: Routledge.

Immenga, U. (1998) 'Basic principles for an international antitrust code', in Hope, E., and P. Maeleng (1998), pp. 46–56.

Kane, H. (1993) 'Managing through prices, managing despite prices', in Zaelke *et al.* (1993), pp. 57–67.

Samuelson, P. A. (1962) 'The gains from international trade once gain', *Economic Journal*, 72, 820–9.

Thrupp, L. A. (1995) *Bitter Harvests for Global Supermarkets – Challenges in Latin America's Agricultural Export Boom*, World Resources Institute (1999), p. 43.

World Resources Institute (1999) *World Resources 1998–99*, New York: Oxford University Press.

Zaelke, D., R. F. Housman, and P. Orbuch (eds) (1993) *Trade and the Environment*, Washington, DC: Island Press.

Zampetti, A. and P. Sauve (1996) 'Onwards to Singapore – The international contestability of markets and the new trade agenda', *The World Economy*, 19.3, 333–44.

Index